W9-ACX-104

The Last Days of Mankind

ECOLOGICAL SURVIVAL OR EXTINCTION

by Samuel Mines

Simon and Schuster
New York

Published by Simon and Schuster
Rockefeller Center, 630 Fifth Avenue
New York, New York 10020

First Printing

SBN 671-20888-8
Library of Congress Catalog Card Number: 73-139645
Designed by Irving Perkins
Manufactured in the United States of America
By Kingsport Press, Kingsport, Tennessee

This book is dedicated to the magnificent men and women of the conservation movement who for so long have selflessly fought for all the generations to come.

The labor of research and of checking the manuscript through its many stages of production was shared by my wife, Susan, whose immense help is gratefully acknowledged.

Contents

Ecology: An Introduction 9

PART I: *The Waters and Wetlands*

1. Rivers, Dams and Floods 27
2. Water Pollution and the Great Water Shortage 49
3. The Everglades 66

PART II: *The Woodlands*

4. The Forests 91
5. Drama in the Redwoods 116

PART III: *The Wildlife*

6. Our Vanishing Wildlife 135

PART IV: *The Blight of Civilization*

7. The Ecology of the City 167
8. The Road Builders 191
9. Man and Pesticides 215

PART V: *In the Public Domain*

10. Public Lands and Public Parks 243
11. Protection or Multiple Use? 263
12. Life in the Balance 291

Index 301

ECOLOGY: An Introduction

THE simplest definition of ecology is John Muir's: Pick up anything and you will find it is hitched to everything else in the universe. Corollary: Disturb anything at your own peril.

Today's conservationists are ecologists rather than nature lovers after Rousseau, branding civilization and all its works as evil. Some conservationists are not naturalists at all, and consider such a label to be more hindrance than help.

One conservation writer admits he is not a bird watcher, is alarmed by women asking him about the cultivation of flowers, and has no interest in pursuing game or fish. His concern is entirely with the destruction of the environment, the problems of urban blight, of slums and pollution. He considers "ecology" to be a more accurate, if less generally understood, word than "conservation."

Charles Darwin, in a rare moment, provided a whimsical slant on ecology in *Origin of Species.* To what, asked Darwin, does England owe her mastery of the seas? Answer: To England's cats. It goes like this:

Red clover flourishes in English meadows. It flourishes because of the prevalence of bumblebees. The bees have long tongues uniquely adapted for gathering honey from clover flowers. In so doing, the bees pollinate the flowers. There are more bees close to the villages. This is because field mice, also in the meadow, which eat the bees' larvae and destroy the honeycombs, are less numerous near human habitation.

The mice, said Darwin, were less numerous near villages because there were more cats where humans live. Now, as cattle feed on red clover and British beef was a staple food of the Royal Navy, it is clear that the cats were responsible for Britain's commanding position as a world maritime power.

Darwin ended it there, but scientist Thomas Huxley, who did more to popularize Darwin than anyone else, carried it further. He said it was generally known that the greatest defenders and collectors of cats were old maids. The real source of England's sea power, therefore, was her old maids.

There is more than a kernel of wry truth in most humor, and in a very exaggerated way the story hints at the intricate web of life that supports us all. Consider the Egyptians.

A backward country, Egypt was understandably proud of the impressive Aswan High Dam, built with the aid of Russian engineers and money to control the floodwaters of the Nile.

Each year, as the rains fell and the snow melted on the slopes of the mountains, the great African watershed was flushed and the Nile overflowed its banks. The dam was created to end this flooding and to store the water by forming the huge Nasser Lake upstream as a permanent reservoir.

That much of the program worked out, but with an unpredicted effect. The flood had brought with it millions of pounds of natural phosphates and nitrates and organic matter to revitalize the fields upon which Egyptian agriculture depended. This was no longer happening. As with all dams, the vital nutrients and a great deal of silt were beginning their long inevitable pileup behind the dam, where they served no purpose except to begin the filling up of the lake.

Previously a great deal of these nutrient-laden waters had rushed out to the Mediterranean, where they had caused a population explosion amongst the plankton—the minute sea plants and animals at the bottom of the food pyramid. The plankton are eaten by small fish, which are eaten by bigger fish, which are eaten by still bigger fish, which are eaten by man. The plankton explosion triggered an annual prosperity in fish all along the chain, the ultimate benefitors of which were the fishermen of the Mediterranean.

With the dam in place, this bounteous overflow was ended. By the fall of 1965, as water backed up behind the Aswan Dam, the flow of the Nile had been so reduced that the plankton languished, the fish in the Mediterranean did not experience their usual boom, and the fishing fleets were in trouble.

Both the food supply and the economy of the Mediterranean countries are affected. Israel is a major loser, having invested heavily in fishing boats and canning factories, but Egypt, Lebanon and Syria are also hard hit.

There is another ecological result. The silt and other sediment once swept out into the Mediterranean by the Nile current shaped the coastlines and upholstered the sea bottom, providing a habitable environment for mollusks, shrimp, lobsters and other bottom creatures. With the current now throttled, the flow of sediment reduced, marine biologists expect the normal currents in the Mediterranean to sweep the bottom bare, destroying this biome and gnawing away at the coastline, to change the landscape and reduce the normal sea life of this area. A biome is specific area in which climate or other factors dictate the predominance of certain forms of plant and animal species.

A fourth result is in the making. The reduction of freshwater flow from the Nile is having an effect on that biological catalyst known as the Suez Canal.

The canal is actually an artificial link between two major marine zones: the Mediterranean-Atlantic and the Red Sea-Indian-Pacific oceans. Barriers in the canal, in the form of three lakes, restricted the movement of marine life from the Red Sea into the Mediterranean. Two of these lakes were so heavy with salt that fish would not penetrate them. The third was brackish enough to be equally impassable. Still, some movement of water from the Red Sea tended to flush out the two saline lakes somewhat. The change in flow from the Nile is expected to have the same effect on the brackish lake. With this flushing action speeded up, marine life from the Pacific may begin to move through the canal into the Mediterranean and ultimately into the Atlantic, changing the characters of both zones.

In fact, Professor Lamont C. Cole of Cornell considers the Aswan Dam Egypt's ultimate disaster. In arid regions, where evaporation is greater than rainfall, the soil becomes more and more saline. Evaporation draws up from the earth only the pure water, leaving the salts behind. Irrigation adds to the process, bringing more water to evaporate, with more salts left behind. Eventually the salinity becomes so severe that nothing will grow and the land becomes desert. All around the Mediterranean, in China and in India, the sites of ancient agrarian cultures, the land is dry and lifeless, although the ancient irrigation ditches are still there. If the Aswan Dam brings a million acres of land under irrigation, as promised, the deadly process will have begun for Egypt.

The story is one of hundreds that might be told. In Borneo a public health team set out to improve sanitation in the villages. One of its projects was to fog the native huts with DDT to reduce the fly and mosquito plague. A short time later the thatched roofs fell in on the occupants, which annoyed them more than the flies had done. It seems that the DDT had killed the wasps that preyed on the caterpillars that lived in the thatch. Unchecked by the wasps, the caterpillars gnawed away the roof bastions in short order. Simultaneously the cats died. They had been catching and eating geckos—small lizards that also lived in the huts. The geckos had ingested enough DDT to make them lethal as food. With the cats dead, the rats moved in, and the villagers now had problems that made them wish they had their flies back again. The public health department sent out a hasty call for cats, which were parachuted into Borneo to fight the rat invasion.

Experiences like these—and worse—illuminate the basic problem. Man, looking upon nature as an adversary, has tried to control and alter his environment to make it more livable for his species. He has done so too often without regard for the ecological havoc being wrought; without understanding that in altering the balances he may be endangering his own existence.

The longshoreman philosopher, Eric Hoffer, no nature lover, argues that humanization means breaking away from nature. Man became what he is, says Hoffer, not with the aid of, but in spite of nature. The argument has point. Only a woolly-minded nature lover who sees the world as a Disney-like, gentle, beautiful meadow peopled exclusively by butterflies, songbirds and adorable bunnies could dismiss it. The meadow is, in fact, a jungle of fang and claw in which every creature lives by killing and eating another. Man has had to alter his environment, to fight heat and cold, storm and drought, disease, famine and insect assault. Nature is as brutal as it is beautiful. Yet man is part of nature and cannot escape himself. As another conservationist, Aldo Leopold, has said, conservation is a state of harmony between men and land.

Nature is brutal because nature promotes only the survival of the species, not of the individual. Man pays at least lip service to the worth and dignity of the individual. Nature wipes out millions of creatures without concern. It takes an Albert Schweitzer to say, "If I am to expect others to respect my life, then I must respect the life I see, no matter how strange it may seem to mine."

Ecology deals with the relations of living creatures to each other and to their environment. The word is derived from two Greek words meaning "study of the home." The word "home" used this way has a more absolute meaning than in the ordinary sense of a family residence. The environmental home is one that an organism may leave only at its peril, since adaptation has fitted it for life under such conditions.

All the living things inhabiting such a home are interdependent, even though they are differently constructed, eat different foods and operate in widely diverse ways. The forest deer is no less dependent upon the earthworm than is the mole, and all three depend upon the soil as they simultaneously draw upon it and help to maintain and create it.

Life on our planet is confined to a narrow band called the biosphere. This band is less than 12 miles in thickness vertically, figuring from the top of the highest mountain to the deepest part of the ocean. Actually, life is not prolific at these extremes. Most marine life is found near the surface of the oceans. On land, life forms are most active from the treetops to just a few feet under the surface of the ground. Within this narrow ring around the planet more than a million different forms of life exist.

The biosphere is divided into biomes, already defined as areas in which soil and climate dictate the kind of vegetation and animal life occurring there. Life in a biome achieves a state of equilibrium in which it tends to sustain itself if undisturbed.

The eastern woodlands of the United States supported their own forest population of animals, birds and insects, which was different from the life of the western plains. The white-tailed deer is a deer of the woods, the red and gray squirrels are tree squirrels. A thousand miles west, on the plains, deer yielded to antelope and bison. The dominating squirrel was a ground squirrel, the prairie dog, which had no trees to climb and had adapted to living in the earth.

It is important to note that these major differences of plant and animal life were dictated by climate and weather. The grasslands of the prairie states are there because the prevailing westerly winds are cooled and lose their moisture as they rise above the Rockies before sweeping down to the plains. With less rain, the drier climate thus produced is not good for trees but is tolerated by grass, and this in turn contributes to the rise of large numbers of grazing animals.

Just below the arctic tundra, across the northern part of the North American continent, stretches a wide belt of evergreens, which are there because of a certain combination of temperature, soil and rain. The mountains of the East and West produce their own biomes out of altitude and exposure. And the southwestern deserts, from Nevada into Mexico, support an astonishingly prolific life well adapted to those arid conditions.

These biomes may be subdivided into smaller characteristic areas. The redwood forests of northern California are nourished on the fog-laden winds from the Pacific and are different from the drier ponderosa forests to the east or the giant firs to the north.

The eastern woodlands vary considerably from north to south, although they include both conifers and deciduous trees. The spruce and firs of the Adirondacks changed to the hemlocks of the Catskills and to the pines and cypress of the South. In the North, deciduous trees run to oaks, chestnuts, maples, elms and birch, with curious little subdivisions like the scrub oaks of eastern Long Island and the pine barrens of southern New Jersey, varying to the oak and hickory forests of the Ozarks.

With each variation comes a different population of animal, bird and insect life. But similar biomes don't always have similar life forms. Polar bears are found only in the Arctic and penguins only in the Antarctic. The climate and general conditions of both frigid areas are not dissimilar, yet neither life form has been able to cross into the other's domain.

There are physical barriers that seal off one area from another. Animals may be unable to cross mountain ranges or oceans, and this keeps an ecological fence around global areas. It is thus possible to delineate six major zoological areas, each with life forms that are different from the others.

The Nearctic realm stretches across the northern part of the globe, taking in the United States, Canada, Europe and Russia. The United States is more like Europe and Asia than it is like South America, even though it is connected to South America by a land bridge and is not connected to Europe. However, it was presumably once connected to Asia by land across the Bering Strait.

In South America we have a Neotropical realm whose animal life includes such tropical forms as monkeys and sloths. South America is more like Africa, where the simian forms add the great apes and where the cats come bigger and are topped by still bigger forms like the elephant and rhinoceros and hippopotamus, which typify the African, or Ethiopian, realm.

The Oriental realm is part of a wide tropical belt, but is different from either the Ethiopian or Australian realms. The Indian tiger and Indian elephant of the Oriental realm are distinctive. And the Australian realm contains several completely unique creatures found nowhere else—the koala, kangaroo, platypus and kiwi.

The concept of land realms was worked out by Alfred Russel Wallace in 1876 and is generally accepted as zoological partitioning of the planet according to animal distribution, although Wallace's original sharp boundary divisions have been much blurred by the importation of many species into new realms.

Whatever the particular characteristics of any region, the basic pattern of ecological dependence remains the same. All the creatures in any biome are part of a great pyramid of life based upon food supply.

On land the pyramid's broad base is composed of green plants, the only living forms that can make their own food. With the magic substance chlorophyll they convert water from the soil and carbon dioxide from the air into starch and protein, requiring only the energy of sunlight. Plants are eaten by herbivorous animals, which convert plant tissues into the proteins of their bodies. These animals are eaten by carnivores, whose digestive systems incorporate the proteins of meat more directly into their bodies. And when any form of life—plant or animal—dies, its tissues are broken down by fungi and bacteria into the basic chemicals, which return to the soil and begin the cycle over again.

One can develop many kinds of food chains. A caterpillar eats a leaf, a bird eats the caterpillar, a snake catches the bird, the snake is eaten by a fox or raccoon—and so on. The essential point is that it begins with the leaf; without the leaf none of the succeeding life forms would survive.

A classic food pyramid is developed in the sea, where plankton form the broad base. Plankton include the green plants of the ocean, microscopically tiny but present in astronomical numbers and equipped with the chlorophyll that through photosynthesis converts basic chemicals into starches and proteins.

The diatoms, desmids and algae of the plankton are eaten by tiny animals, some of which are themselves classed as animal plankton: minute shrimplike copepods and worms and the larvae of various mollusks and crustacea, and tiny fish. These are the prey of larger fish, such as the prolific herring. And these are eaten by still larger fish or seagoing animals that are hunted for food by man.

At each step of this pyramid there is a ten-to-one loss. For a thousand pounds of plankton there will be produced about a hundred pounds of animal plankton, which will nourish ten pounds of fish, which, consumed by man, will enable him to generate one pound of body weight. This ratio is paralleled by the diminishing numbers of the pyramid. The astronomical numbers of plankton are succeeded by a smaller number of animals that feed on them, and these by a still smaller number of fish that eat them, and so on up to the relatively few humans at the apex of the pyramid.

The interdependence of living creatures has more ramifications, however, than the simple one of eating each other. Patterns of partnership or competition have evolved on many levels. Mutualism is a relationship beneficial to both without harm to either, as between bees and flowers. The flowers offer food in the form of honey and pollen; the bees perform the act of pollination for the plants.

Certain soil bacteria affix themselves to the roots of leguminous plants and, the bacterial cell being a remarkably efficient little chemical factory, extract nitrogen from their surroundings and make it available in a form the plant can use.

The hermit crab, possessing no sturdy shell of its own, borrows a discarded snail shell, then often plants a sea anemone on its back. The anemone, with its stinging tentacles, provides police protection, and gains transportation as the crab moves about.

A classic partnership in the microscopic world is the symbiosis between the hydra, an animal, and the alga, a plant. The alga lives in the digestive tract of the hydra and gains shelter, plus some food substances. In return it supplies oxygen and the starch-protein results of photosynthesis.

No less important are the various bacteria that live normally and harmlessly in the intestine of man, utilizing some of the food, but playing a major role in digestion. Similarly, a particular protozoan lives in the digestive tract of the termite and without it the termite would be unable to digest the wood it eats.

Harmless and beneficent partnerships shade off; there are vines, like the mistletoe, that do some harm to the trees on which they grow; there are other parasites that cause extensive injury and death to the host. These occur in both the plant and animal worlds.

Fungi, which are everywhere, perform an invaluable service in breaking down the chemical structures of dead plant life so that new growth will be nourished by the vital elements returned to the soil. Without fungi the forest floor would pile up endlessly with trunks, stems and leaves that would not decay. Some fungi, like mildew, cause disease in living plants, and there are fungus diseases of animals and man as well.

Animal parasites are legion, from the louse to the tapeworm. All cause injury to the host, not only by sucking its blood but often because the parasite carries its own parasites in the form of bacteria, protozoa, or viruses, which may cause serious disease in the host.

The virus is the ultimate parasite. Possessing no metabolic functions of its own, it does not eat or show other signs of life—except one. It can penetrate a living cell, and once inside, direct the cell's control center, the DNA, to produce more viruses—a process that can result either in the death of the cell or its conversion to a malignant form.

From the objective point of view, the parasite is no worse than any other predator. Certainly it is no worse than man, who raises chickens for eggs and rewards them by killing and eating them as soon as their egg production ceases. Or breeds cows for milk and caps their careers with the same ignominious trip to the butcher.

Man, however, persists in seeing things from an anthropocentric point of view. There is, for example, a campaign by sportsmen's groups in this country to "gang up on killer crow." It appears that the common crow has earned the enmity of hunters by raiding ducks' nests. Crows, it is alleged, like eggs—as do humans. Sportsmen are therefore exhorted to hunt and kill crows, "whose business," they claim, "is murder."

These sportsmen see nothing ludicrous in the fact that their passion to save ducks is not for the love of ducks, but only to kill the ducks themselves. It obviously never occurs to them that, from a purely objective point of view, the crows have possibly more rights in the matter than humans, as crows hunt only for food, while humans kill for sport. Nor should we overlook the fact that the ducks would undoubtedly have a very different point of view about the whole thing.

It is an axiom in ecology that a natural predator never endangers the existence of a species. The existence of a predator is a necessity in nature's precarious balance. Predators prevent overpopulation, eliminate the weak and sickly, and thus upgrade the species by ensuring the survival of the strong.

Only man, with vastly improved means of destruction, has upset the predator balance. No multiplication of wolf packs could have virtually wiped out the American bison, but commercial hunters did it rather quickly, the kill running as high as 250,000 animals a month. General Phil Sheridan, who, after the Civil War, was military commander of the Southwest, encouraged the slaughter as the best way of crippling the Indian tribes, which depended upon the bison for food and hides.

Buffalo Bill Cody, who made his living by shooting bison for railroad work crews, claimed 4,280 animals killed by himself alone. Another gun-happy celebrity of the day was Billy Tilghman, part-time sheriff and hunter, who claimed to have killed 3,300 bison in seven months. The buffalo hunters left great piles of bones and rotting carcasses on the prairie, since many of them killed only for the hides.

From several points of view, man is both predator and parasite.

But man's destructiveness is not limited to these roles alone, for it includes pollution, deforestation, soil erosion and other upsets of the ecological balance. Man introduced rabbits into Australia, the Norway rat into the United States, starlings and English sparrows into New York; brought the Russian thistle to the West and the water hyacinth from Venezuela to Florida, where it now clogs rivers and canals, the Japanese beetle and the Mediterranean fruit fly to America—the list is almost endless. There is even a species of catfish that has managed to get from Africa to Florida, a catfish that leaves the water and walks around on dry land, tangling with dogs and scaring the citizenry speechless. These imports tend to run wild because of the absence of natural predators that controlled them in their original environments.

The case of the sea lamprey well illustrates the clumsy hand of man. If you look at a map of North America you will see that of all the Great Lakes at the top of the United States only Lake Ontario has a connection to the Atlantic Ocean, by way of the St. Lawrence River.

Into Lake Ontario from the Atlantic came occasional sea lampreys. The lamprey is a parasite. It is an eel-shaped creature looking rather like a piece of garden hose about a foot and a half in length. Up front is a circular mouth like a suction cup, rimmed all around with small, sharp, curved teeth. The lamprey's goal in life is single-minded and simple. It attaches itself to an available fish of a good size. The suction cup holds, the teeth rasp through the fish's skin like a circular saw, and the lamprey begins a workless idyll of sucking its host's vital juices until the fish finally weakens and dies. Whereupon the lamprey goes in search of another victim.

Lampreys were no great menace in Lake Ontario, where they did not proliferate because they lacked the deep cold water they needed and the right kind of inlet streams for spawning. Although they adapted easily to fresh water, they could not get past Lake Ontario, barred from further inland penetration by Niagara Falls.

In 1829 the Welland Canal was built to bring ships around the Falls into Lake Erie. The early canal was too small to make much difference, but in the period between 1913 and 1918 it was enlarged and deepened. Presently lampreys were found in Lake Erie. Because of conditions similar to those in Lake Ontario, they did not flourish here either, and no one was alarmed.

It took many years for the lampreys to filter slowly northward into Lake Huron. And now they found what they needed: deep cold water, small inlet streams for spawning and a veritable banquet of food—lake trout, whitefish, burbot, bass and ciscoes. They had, in fact, landed in one of the important commercial fishing areas of two countries, the United States and Canada.

In just about twenty years the fishing industry of the Great Lakes was all but wiped out. The catch of lake trout from Michigan and Huron dropped from 11 million pounds a year to a few hundred pounds. Boats rotted on the beach, nets flapped untended in the breeze. Lakeside villages were practically ghost towns. By 1955 the situation was regarded as all but hopeless.

Since 1941, however, two biologists from the Michigan Department of Conservation, Vernon C. Applegate and John H. Howell, had been searching for an antilamprey defense. They had tried, and discarded, fencelike weirs across the streams to block the spawning migration—even electrically charged weirs. They next began to screen chemical compounds for one that might selectively poison the lamprey larvae during their long growth period in the mud bottom, without harming other fish.

Six years of work and disappointment and fund-cutting crises went by before they had it—a chemical of a halogen-nitrophenol group called TFM. It killed the young lampreys with "very little effect" on other fish. A two-year test proved successful, and by 1960 a cooperative effort between the United States and Canada saw streams on both sides of Lake Superior treated. The spring of 1962 brought a collapse of the lamprey population to about a seventh of its former numbers.

But an ecological balance is not so easily restored. The disappearance of trout and whitefish from the lakes left room for another invasion from the sea—millions of alewives, little silvery fish that rapidly filled the lakes, ate the eggs of other fish and periodically died by the ton on the beaches, adding a new kind of pollution.

Another poison was out of the question. So the conservation departments began thinking in terms of predators. Work had already begun on hatching lake trout to replace those destroyed by the lampreys. For predators that actually liked an alewife diet the biologists chose two species of salmon from the Pacific Coast, the

Chinook and the coho. Both are big, active fish, highly prized for sport and food, and both would eat alewives.

The results topped anyone's expectations. The salmon took to the lakes with enthusiasm and their population boomed. The ghost towns came to life and fishermen became delirious at the prospect of hooking 20-pound fish that give them the fight of their lives. In 1967 an estimated 150,000 fishermen went after coho in Lake Michigan and its tributaries alone.

So now only a few problems remain. How long will the alewives last as food for hordes of big, hungry salmon? How will this import affect the rest of the fish population in the lakes? And what can be done about the DDT washing into the lakes from the farmlands above, which is already at a dangerous level in the eggs and young of the salmon?

Our understanding of ecology is still precariously slim, so slim that many states are still offering bounties for the killing of such natural predators as cougars and coyotes. Even the conservation departments are slow to realize that predators fill an essential role in the life chain and that man upsets it at his peril.

Some years ago in Canada there was an outcry that wolves were killing off the caribou, and the northern provinces demanded bounties on wolves. A biologist was sent into the Yukon for a two-year study on the life habits of the timber wolf. Actually, although he did not know it, his mission was to bring back proof of the wolves' guilt so that an appropriation might be requested for the bounties.

The biologist did not fulfill expectations. After two years of living in proximity to several wolf families he produced a report which stated bluntly that wolves were not decimating the caribou at all. The major part of the wolves' diet, he said, was field mice. Only occasionally did they take a caribou and then it was usually an old or sick one whose removal did the herd far more good than harm. The wolf not only benefited the caribou herds but was essential in rodent control.

Moreover, he reported, legends of the wolf's fearsome character were largely fiction. Wolves were not savage or dangerous, as portrayed; they were even-tempered and playful, family-minded and loyal. They had, in fact, most of the characteristics of man's best friend, the dog.

As to the real menace to the caribou, it came from the very people who were screaming for the bounties on wolves—the prospectors and hunters. They were shooting caribou so lavishly to feed their large and hungry dog teams that for miles around every cabin in the North there was a perimeter of bleaching caribou bones.

Needless to say, the biologist was promptly fired. He hadn't brought back the report he was sent to get.

Much the same situation has been true in the United States, where for years a ruthless war has been waged against prairie dogs, coyotes, foxes, wolves, cougars and hawks. In 1965 some $5,575,000 in Federal funds was spent in predator "control." Poison bait was scattered lavishly to kill prairie dogs, wolves, foxes and coyotes. Eagles, hawks, crows and vultures feeding on the carcasses died as well. Killing the predators caused the usual population explosion among the small rodents such as mice, and among insects, with consequent damage to the grasslands.

There is ample evidence to question the entire concept of mass assault on any predator or supposed enemy. Whole forests are sprayed to control the gypsy moth or the tent caterpillar. What usually happens is that the spray kills many of them, giving temporary benefit, but those resistant to the poison are spared to breed a new race of immune insects. Moreover, the spray kills the natural predators and parasites of the insects, thus reducing biological controls.

Experience reported by French forestry stations just after World War II adds practical confirmation. The Carnac pine forest in Brittany was sprayed with DDT by helicopter in 1949 to control caterpillars. First results were good, but later there was a multiplication of resistant types that could no longer be controlled by DDT. As the natural predators were also destroyed, keeping the caterpillars in check now became a serious problem.

In 1956 biological controls were introduced in the Mont Ventoux area. A virus parasite of the caterpillar was found and the crawlers virtually eliminated in two years. However, this raises a question. Is the elimination of a life form, even a caterpillar, ecologically sound? Because we class a leaf-chewing caterpillar as a pest, does this mean it has no place in the life chain?

In the summer of 1966 a heavy invasion of tent caterpillars hit the Prince Albert National Park in Saskatchewan. Millions of caterpillars swarmed through the forest and stripped the trees and bushes as bare as telephone poles. The trees were covered with wriggling masses of hairy caterpillars and the forest was noisy with the crunching of leaves.

When they had gone through, the forest appeared dead. But it wasn't. Three weeks later small leaves began to appear. They grew out, but never attained the size of the original leaves.

Meantime the caterpillars had gone into the pupa stage. But their very numbers had created a trap for them. A parasite fly had found its own bounty, and millions of pupa cases had an egg implanted from this fly. The fly egg hatched early, before the metamorphosis of the pupa began, and the larva fed on the still dormant pupa, which never completed its mission to become a moth. So while the forest was recovering, nature was asserting its balance in caterpillars, with the promise that the next year would see fewer of them.

But the strangest part is this: the park had had very little rain and the forest was quite dry. When the caterpillars stripped the trees of leaves they checked the normal transpiration (the exhalation of water through the leaves) that goes on continuously. Normal transpiration would have been at a rate that would have seriously depleted the trees in this dry spell; by checking it, the trees got by on much less water. In short, the caterpillars actually *saved* the trees.

The new leaves that came in a month later were small enough to keep the metabolic machinery going on a reduced scale. By autumn it was noticeable that the area that had been stripped by the caterpillars was now greener than the rest of the forest, which suffered drought damage. Nature had again restored the balance.

Carl Reidel, of the University of Minnesota School of Forestry, comments (*American Forests*, Feb. 1968) that a forest or a lake is an interlocking system of life so staggeringly complex that we have only a bare inkling of its mysteries. In conservation, he says, we consistently practice brinkmanship—the art of waiting until almost too late. We do nothing until a crisis is upon us, then we mount a crash program, deal hastily and superficially with a problem that is hardly solved, sometimes made worse. There is even reason to fear that in many matters it may already be too late.

Professor Lamont C. Cole of Cornell University asks the ominous question: Can the world be saved? He believes the destruction of our environment has already passed the point of no return. As a simple illustration: A jet airliner of the Boeing 707 class burns a ton of kerosene about every ten minutes. By-products of this combustion are about one and a third tons of water and nearly three tons of carbon dioxide. About 10,000 airplanes a week land in New York alone from various parts of the country. If we assume the average flight to be about four hours, this would amount to some 36 million tons a year of carbon dioxide released into the atmosphere from jet flights terminating in New York alone.

Jet planes are not our greatest source of carbon dioxide. Combustion from hundreds of sources—heating, generating power, incinerators, and so on—is pumping more and more carbon dioxide into the atmosphere each year. Meanwhile man is busily spreading over the earth and removing more and more green plants as he adds concrete and paving. Professor Cole estimates that a million acres of greenery is lost every year. This a major source of oxygen. Another source, water-borne plankton, is being crippled by spreading pollution of our water supplies.

Thus we seem to be losing the battle of our oxygen supply, which is usually fatal. An effect that may come earlier is the "greenhouse" effect. The increased carbon dioxide tends to raise the earth's temperature, with all kinds of interesting results, such as melting the polar ice caps—which would raise the ocean level about 300 feet and wash out most of the major cities of the world.

The race is now between man's blind capacity for destruction and the growth of our understanding of the great web of life—ecology. There is no indication that we are winning that race.

PART I

The Waters and Wetlands

CHAPTER

I

Rivers, Dams and Floods

EVERY living creature affects and alters the ecosystem in which it lives and so affects every other creature. The great web of life covers not only the present but extends backward into the past and inexorably forward into the future.

This theme was popular a few years ago with science-fiction writers. A time traveler goes far back into the past. He is told he may observe, but is warned not to touch anything. He is very careful, but, unseeing, steps on an ant. Millions of years in the future a mountain range, three lakes and half a dozen cities vanish, and the man himself disappears, his entire family never having been born.

So every action affects both its own time and the future. Consider the beaver, a compulsive worker and an engineer by instinct. He finds a small stream and dams it, backing up the water to form a pond. For the beaver the pond has two essential uses. It is his natural habitat in which he builds his home, a lodge of sticks and mud. And it enables him to cut and float along the shore the trees whose bark is his food and whose branches are his building material.

The trees along the original stream are, of course, drowned; they are promptly taken over by woodpeckers, which hollow out cavities that provide homes for themselves and other birds, including some species of ducks, such as the wood duck.

The beaver pond stabilizes the water supply and tempers the immediate local climate; it furnishes a drinking area for wildlife and a refuge for waterfowl. As the beavers use up the trees along the shoreline, they must go deeper into the forest for food; and since overland travel is not to their liking, nor can they drag trees over the ground, they dig canals, often for some distances, into the surrounding woods, further affecting the local ecology.

Except for the few trees consumed or drowned, everyone seems to benefit. Dams therefore appear to be a good idea, an idea bought with enthusiasm by the U.S. Army Corps of Engineers and the Bureau of Reclamation. From the list of dams proposed by them it would seem that every stream in the United States is slated for at least one, and some for many.

Unhappily there is a large difference between a small beaver dam that transforms a marshy spot into a pond and a giant concrete structure, hundreds of feet high, erected in a spot too often chosen for political rather than ecological reasons.

There are few better illustrations than the curious sequence of events surrounding the most notorious of recent dam episodes, the Bridge Canyon and Marble Gorge dams proposed for the Colorado River in the vicinity of the Grand Canyon.

The business started with something called the Pacific Southwest Water Plan, designed to bring water to Arizona and southern California. The plan was made up of various parts including Bridge Canyon Dam at $511,000,000; Marble Gorge Dam at $239,000,000; the California Aqueduct Enlargement, to bring water from northern California to southern California, at $240,000,000; and the Central Arizona Water Project at $527,000,000. A follow-up project, to build dams and reservoirs on certain rivers in northern California and bring the water south by aqueduct, was blue-printed at $1,422,-000,000, but that was to come later.

The heart of the Pacific Southwest Water Plan was the Central Arizona Water Project, and this grew out of Arizona's struggle with California for its share of Colorado River water. California, it seems, was taking more than its share, and in 1952 Arizona filed suit against California in the Supreme Court. The decision was eleven years in coming; in the meantime southern California's growth and water demands continued as though some celestial advocate had whispered assurance that Angelenos could go on drawing the lion's share of water from the Colorado forever.

But in June 1963 the Supreme Court lowered the bucket. California would be entitled to 4,400,000 acre-feet of water a year and Arizona to 2,800,000 acre-feet a year. An acre-foot is the amount of water required to cover one acre of ground to a depth of one foot. Nevada was allowed 300,000 acre-feet, making a total of 7,500,000 acre-feet for the Lower Colorado Basin states. Another 1,500,000 acre-feet was guaranteed Mexico by treaty.

The decision didn't sit too well with the upper basin states of Colorado, Wyoming, New Mexico and Utah, which preferred a compromise bill sponsored by Congressman Wayne Aspinall of Colorado calling for five upper basin dams costing $360,000,000.

Arizona's problem lies in the fact that Phoenix and Tucson, its two principal cities, are in the desert. Tucson has neither river nor lake—no sources of surface water at all. The city pumps all its water from underground and is lowering this underground water table by three feet a year. Phoenix, larger than Tucson, draws partly on the Salt River, but it is lowering its own water table by eight feet a year. Thus Arizona's compulsion to push the Central Arizona Project and the two new dams at Marble Gorge and Bridge Canyon. The river was already well dammed, with Headgate Rock, Palo Verde, Hoover, Parker Davis and Imperial dams, all with reservoirs, each turning its section of river into a placid lake.

Bridge Canyon would be the highest dam in the western hemisphere, 740 feet high, located downstream from Grand Canyon National Park. It would raise the river level at the dam by 700 feet and back the water up 100 miles, flooding Grand Canyon National Monument and continuing 13 miles into Grand Canyon National Park.

The second dam, Marble Gorge, was to be located above Grand Canyon National Park. The string of reservoirs projected behind this dam would drown the river all the way back to Glen Canyon, leaving only104 miles of river out of 1,000 miles in its original state.

But before either of these dams was built, something had to be done. Through eons the Colorado River had carved out the fantastic sculpture of the Grand Canyon because its waters carry a heavy load of sand, silt and stone—natural chisels to work the sandstone of these ancient cliffs. With such a burden of sediment backing up behind the new dams, silting would be a problem. The Bureau of Reclamation came to the rescue by proposing a dam farther upstream, at Glen Canyon. This would cut the pace of the flow and trap much of the silt before it reached Marble Gorge. Moreover, Glen Canyon Dam could be justified on the basis that it would generate hydroelectric power, form Lake Powell as a recreation area and reservoir, and control flood stages in the river.

Glen Canyon contained some of the most beautiful caverns and rock formations along the river, and conservationists were appalled. But they were also caught off guard, rallied too late, and lost. Glen Canyon Dam was completed in 1963, drowning completely one of the most magnificent canyons of the Colorado gorge and seriously impairing the famous Rainbow Bridge.

The gates of Glen Canyon Dam were closed early in 1963 and the long backup began to fill Lake Powell. Below, the flow from the river was cut to a thousand cubic feet a second, posing a problem to the turbines downstream at Hoover Dam.

In March 1964 Lake Powell contained only 3,000,000 acre-feet of water, half the amount needed to begin the generation of electricity through its own turbines. At the same time the level of Lake Mead behind Hoover Dam had dropped to 1,123 feet—the critical level below which power generation would stop. Secretary of the Interior Stewart Udall ordered the Glen Canyon Dam gates opened to save Lake Mead, touching off a storm of protest from the states in the upper basin. Unsympathetic conservationists predicted darkly that given a few dry years, it could take twenty years to fill Lake Powell.

Samuel B. Nelson, chief engineer of the Los Angeles City Department of Water and Power said: "To reduce production of 18 electrical generators at Hoover Dam in order to start one at Glen Canyon, and months later another there, and months later a third—maybe—is supposed to be supported by Federal figures, which are yet to be given for checking by engineers who advise other interested public agencies. When we receive these data as now promised, and analyze them, more detailed comment may be made on this financial aspect of the power situation. As of now, we are skeptical." *

* In *Time and the River Flowing: The Grand Canyon* by François Leydet. The Sierra Club, San Francisco, 1964.

In addition, ecologists point out that for a river supposed to supply water, a huge amount—6,100,000 acre-feet—would now be sequestered in dead storage behind Glen Canyon Dam. Moreover, the evaporation from newly created Lake Powell was estimated at 650,000 acre-feet, enough water to supply a city the size of Denver.

Such was the prelude to the next move in the Central Arizona Project, the building of two more dams at Marble Gorge and Bridge Canyon, each with a string of reservoirs behind it. These reservoirs were estimated to lose by evaporation another 100,000 acre-feet of water, or about twice the annual consumption of Tucson.

How, asked the Sierra Club, would the Bureau of Reclamation obtain the water to fill these reservoirs? By emptying Lake Powell again? The Marble Gorge and Bridge Canyon reservoirs would hoard another 4,073,000 acre-feet of water. Together with Lake Powell, this totaled *a year's average flow* of the Colorado.

There was still another part to this program—the Kanab Diversion Project. This was a plan to build a tunnel from Marble Gorge straight west to Kanab Creek, bypassing Grand Canyon National Park and diverting 90 percent of the water to a new set of generators in Kanab Canyon. The result would be to reduce the Colorado through the stretch of Grand Canyon Park to a trickle or a dry stream bed. It wouldn't matter—tourists standing on the South Rim at Bright Angel Lodge could hardly see the river anyway.

The mathematics here were simply that everyone was dividing up more water than there was. The river was being asked to supply 7,500,000 acre-feet a year to the upper basin, as much again to the lower basin, 1,500,000 acre-feet to Mexico under a 1944 treaty, fill the string of reservoirs behind the new dams at Marble Gorge and Bridge Canyon, Lake Powell, Lake Mead, Lake Mojave and Lake Havasu (behind Parker Dam), and keep up with the evaporation of all these new bodies of water, plus the enhanced seepage through the porous rock bed—particularly leaky through the Redwall limestone of Marble Gorge. Bureau of Reclamation engineers themselves estimated a seepage loss at Lake Powell of 12 to 15 percent. They underestimated notably. Present reports indicate that 25 percent of Lake Powell's water is disappearing through the porous rock walls.

Forgetting for a moment the damage to the esthetic and scientific aspects of the Grand Canyon—its incomparable beauty and grandeur, its irreplaceable value as a geological monument, a great vertical slice through time into the youth of our planet—were these dams ever justified from a purely economic point of view?

It was pretty generally admitted that the two new dams would provide no new water supply for Arizona. The real purpose was hydroelectric power. About a quarter of this power would be used to pump water from Lake Havasu, far downstream, to Phoenix and Tucson. The remainder of the power was to be sold at a profit, to pay for building the Marble Gorge and Bridge Canyon dams and to finance the rest of the Central Arizona Project. Optimistic estimates that the electricity would be sold for 6 mills per kilowatt-hour led to the dams' being affectionately known as the Bureau of Reclamation's "cash registers." This enthusiasm did not spread to Senator Clinton Anderson of New Mexico, who pointed out that a new thermal plant being built at Four Corners, where Utah, Arizona, New Mexico and Colorado meet, would be producing electricity for 4 mills per kilowatt-hour. And the U.S. Office of Science and Technology estimated that nuclear plants would be producing power within ten years at 2.5 mills per kilowatt-hour. Such developments would play havoc with the projected amortization of Glen Canyon Dam and the other dams, based on selling power at 6 mills and making them the most costly power projects in the Southwest.

The Sierra Club offered a comparison between the cost of the two new dams, $750,000,000, and $263,000,000 for a steam plant that would produce the same amount of power—a saving of $487,000,000. Moreover, the dams would require long and expensive transmission lines, whereas a steam plant could be built close to its consumers.

Richard Bradley, Associate Professor of Physics at Colorado College, wrote: "Regardless of whether or not nuclear energy ever becomes a reality, hydropower is destined for ever decreasing importance in the United States. It is well known that if every bit of usable stream in the country were developed to its full capacity, the total hydropower generated would furnish only about five per cent of our total energy requirements and these are expected to double in twenty years." *

There are other forces of obsolescence to consider, such as accelerated salinization. The water of the Colorado carries not only grains

* *Pacific Discovery*, journal of the California Academy of Sciences. Quoted in *Time and the River Flowing*.

of sand but a rich load of dissolved salts of calcium, magnesium, sodium, and bicarbonates. Water evaporates much more rapidly from a reservoir than from a running river. When it evaporates, these salts are left behind. Tests made at Hoover Dam in 1956 and 1957 showed 791 parts per million of solids in solution, compared with a top acceptable figure of 500 parts per million for drinking water. In other words, the water in Lake Mead was already too saline for drinking purposes. Additional evaporation from two reservoirs below Hoover Dam—Lake Mojave and Lake Havasu—plus additional brine pumped into the river from a reclamation project in the Wellton-Mohawk Valley in Arizona completed the salting. As the Sierra Club reports, in the winter of 1961 the water reaching the Mexican border showed 2,700 parts per million of salts. The Mexican government claimed crop damage on 100,000 acres of land and diverted the water into the Gulf of California.

When the Mexicans complained to Washington, the State Department replied, incredibly, that the 1944 Treaty promised Mexico 1,500,000 acre-feet of water a year, but didn't guarantee that it would be usable.

On balance, then, the Marble Gorge and Bridge Canyon dams would not conserve water or supply it to Arizona, and would not produce electricity economically. In testimony before the House Interior Committee's Subcommittee on Irrigation and Reclamation a little-known 1963 report of the National Park Service came to light in which Secretary Udall had opposed and condemned the entire dam-building program. He then reversed himself and supported the project, but in 1967 reversed himself again and opposed it, to support a plan to buy power from a confederation of utilities called WEST (Western Energy Supply & Transmission Associates). This, said *Business Week* (April 29, 1967), "irritated traditional reclamationists, including the Bureau of Reclamation, who feel dams are essential to any reclamation project." It also suggested to "some critics" said *Business Week*, not naming them, "an unnecessary capitulation to conservationists." A variety of other plans were proposed, including one from Congressman Craig Hosmer of California (who evidently found little that was endearing in Secretary Udall's attitude) that the Department of the Interior invest instead in a Las Vegas gambling casino as a money-maker.

In the end it was neither logic nor economics that defeated the dams. An awakened conscience perhaps, a growing unease that we didn't know where an aggressive and undisciplined technology was taking us—something of this sort had at last begun a ferment.

That all-but-forgotten Park Service report had said quietly that Bridge Canyon Dam would change the character of a particularly scenic length of wild river to something far less desirable for a national park. The river, with its ever-changing currents, pools and rapids would be blotted out by the slack water of the reservoir. The natural stream-bank ecology would be drastically changed and the existing plant and animal life would be drowned out. Silt and debris accumulation would be inevitable. And, a minor point perhaps, the rise and fall of water in the reservoirs would reveal rock walls once glowing with colors, now slimed and foul with mud.

On February 1, 1967, the government publicly abandoned its plans to build the Colorado dams and proposed instead the construction of a steam plant to generate electricity for pumping water from the lower Colorado to Arizona. Congratulated on his victory, Secretary Udall said dourly that it was a victory for common sense.

If it was a victory, it left two people something less than happy. Congressman Aspinall said the decision was unacceptable. And David Brower, then executive director of the Sierra Club, said he was left with misgivings about a provision for another dam site, Hualapai, left open for future action by Congress. Moreover, there was no assurance, especially in the light of Aspinall's remark, that the Bureau of Reclamation wouldn't try again.

A curious sidelight on dams comes from *The Sciences* (November 1968), published by the New York Academy of Sciences, in an article titled "Dams That Shake the Earth." "Fault areas," says the story, "ruptures or lines of weakness in the crust of the earth are particularly susceptible to earthquakes caused by dam construction. In these regions, faults that have remained inactive for years are unable to stand the stresses imposed by the accumulation of tremendous quantities of water."

During the fifteen years preceding the building of Hoover Dam on the Colorado there were no earth tremors in the area. In September 1936, as the water in Lake Mead reached a depth of 330 feet, the first tremor was recorded. In the following year the water rose to 400 feet, and more than 100 tremors were felt. Seismographs were installed near the dam in 1938, and since then *several thousand* tremors have been recorded.

Geologists estimate the weight of the water in Lake Mead at 35,000,-000,000 tons. This enormous load created rock tensions which were released along fault lines. Known faults north of the dam had been inactive since the Pleistocene era but were activated by the over-load of water.

This has happened in many parts of the world where dams have been built. Near Bombay, India, a serious earthquake occurred fol-lowing the construction of the Koyna Dam. Two hundred people were killed and there was widespread damage in a large area.

The Kariba Dam, built on the Zambesi River in Africa, created one of the largest artificial lakes in the world, and a region that had been seismically inactive began experiencing tremors. Recording began in January 1962. There were 30 tremors in five days of March, and they increased in number until late September, when the area was jolted by a series of major earthquakes. The tremors then de-creased in severity, although small ones continued.

In Greece a dam across the Acheloos River formed Lake Kre-masta. The lake began to fill in July 1965. It was nearly full in Feb-ruary 1966, when a major earthquake occurred. Nearly 500 houses collapsed and more than 2,000 were damaged. Sixty people were in-jured and one was killed.

Still, the dam builders are not quite through. To a certain type of mind a free-flowing, wild river is an affront. It is a waste unless it is harnessed, diverted, culverted; used to flush sewage, to make steel or chemicals, or to cool reactors. The miracle of grace and beauty, the magic of running water, the harmony with nature touches them not at all.

If the Grand Canyon is unmatched in grandeur, the Hudson River is as beautiful in its own way. The Hudson is a mountain stream, plunging out of the Adirondacks from its birthplace in Lake Tear of the Clouds on Mount Marcy. Flowing south past Albany, it by-passes the Catskills and enters the gorge of the Hudson Highlands, twisting its way between mountains rising sheer from the water, with names that come from the lips like poetry—Storm King, Break-neck Mountain, Bear Mountain, Anthony's Nose, Dunderberg Mountain and High Tor. Once through the broad area of Haverstraw Bay, the Hudson flows under the Palisades, sheer rock cliffs 500 feet high (and barely saved from the sand and gravel miners), and then becomes a deep-water fjord, an estuary of the sea, with regular tides. This Hudson River canyon runs far out into the ocean, its great depth providing ideal navigation for the big ships that make New York a major port.

Such a tidal river has no flood-stage problems and seems an unlikely candidate for the dam builders. Yet there are already some twenty dams on the upper reaches of the Hudson. In 1968 the Army Corps of Engineers, with the New York State Water Resources Commission, issued a feasibility study on the building of "Gooley No. 1 Dam" in the Adirondacks, just downstream from the junction of the Hudson and Indian rivers. The purpose was to regulate stream flow and furnish additional water to New York City.

The proposal illustrates an important principle—that no conservation victory is permanent. This dam, estimated to cost $57,-000,000, would require 16,000 acres of land, of which 14,500 would be flooded, forming a reservoir 25 miles long in the middle of the Adirondack preserve, which the state constitution guaranteed to remain forever wild.

The reservoir would eliminate some of our most dramatic eastern forest wilderness. It would inundate the historic village of Newcomb, where Theodore Roosevelt took the oath of office as President of the United States. He had been climbing Mount Marcy when a messenger brought the news of McKinley's assassination, and he made the hurried dash on foot 40 miles to the railroad.

"The upper reaches of the Hudson and its tributaries are wild rivers," says a 1966 report of the Hudson River Valley Commission recommending that they be given this protected status by the legislature. The area is a remnant of the once great eastern forest, an oasis more than ever important in so heavily populated a state as New York. Organizing the opposition is the Adirondack Hudson River Association, which maintains that the Army Engineers have not even tried to find alternate sources of water. The greatest source in New York, the underground supply containing 90 percent of all the state's fresh water, has hardly been tapped. "We are not talking about the loss of a portion of the upper Hudson," says a leaflet of the organization. "We are talking about the destruction of the world-famous Adirondack Park! To drown out Yellowstone National Park would not be more outrageous!" *

Gooley Dam is merely the newest threat to the Hudson. An old antagonist is New York's electric utility company, Consolidated Edison, which proposed to build a 2,000,000-kilowatt nuclear-powered, pumped-storage hydroelectric plant at Storm King Mountain in the town of Cornwall just above Bear Mountain Park.

With heavy-footed subtlety Con Ed announced its plan as "creative conservation" and promised that Storm King would look better after the completed project than before. The landscaping of the shoreline was to compensate for blasting out part of the base of the mountain to install eight turbines, and boring a tunnel 40 feet in diameter on a rising grade up through the mountain and beyond, into a high valley two miles away. The valley contains a fresh-water reservoir for the town of Cornwall, and part of the surrounding woodland is Harvard's Black Rock Forest, an outdoor laboratory of forestry research.

Con Ed proposed to take over this area and build five dams and a dike to convert the valley into a giant storage basin of 8,000,000,000 gallons. During slack hours of demand for electricity, such as at night, water from the Hudson would be pumped up through the tunnel to fill the basin. When power demand was high the dike would be opened and the water would rush back down the tunnel to turn the turbines and generate electricity to meet the peak demand. The power would be transmitted under the Hudson, then sent on high overhead lines across Westchester County to New York City.

This would be a hydroelectric system of sorts. But since it requires more energy to pump a given amount of water uphill—2.5 kilowatts—than is regained when it rushes down again—2 kilowatts—the plant was not expected to show a profit. Its only purpose was to provide peaking power when it was needed.

There were a few problems. First, the Cornwall reservoir. Con Edison proposed to solve that by setting aside $2,000,000 to build Cornwall a new reservoir elsewhere. Then there was the question of

* *The Impending Tragedy of the Upper Hudson Region*, July 25, 1968. The Adirondack Hudson River Association, Burnt Hills, N.Y.

encroachment on Harvard's experimental forest. President Nathan M. Pusey of Harvard made it clear that he opposed any reduction in the size of that property.

Early in 1963 Con Ed applied to the Federal Power Commission for a license to build and announced that fact to its stockholders in its annual report. A New York attorney named L. O. Rothschild received his copy of the annual report and was far from pleased at the prospect. In November the Scenic Hudson Preservation Conference was set up with Rothschild enlisting people like Carl Carmer, Brooks Atkinson, Aaron Copland and David Sive of the Sierra Club. Other conservation groups—the Audubon Society, the Conservation Foundation, the Wilderness Society, the American Forestry Association, the Nature Conservancy, the Adirondack Mountain Club, and others—more than 30, joined the fray.

Scenic Hudson's petition to testify at the Federal Power Commission hearings was denied. A special hearing obtained in Washington had no effect. On March 9, 1965, the FPC approved the Con Edison application and granted the license. Scenic Hudson moved now to the U.S. Court of Appeals, which, on December 29, reversed the FPC decision and canceled the license to build. It did so in rather blunt language, telling the FPC in so many words that it had failed in its role as protector of the public interest.

The decision, in fact, was a precedent-making one, for it recognized the right of Scenic Hudson to sue as an aggrieved party in the public interest, even though the group itself had suffered no economic loss. This is a principle we shall be hearing of again as more and more conservation groups utilize their right to sue in public courts as representatives of the public interest against privileged groups. For the first time protection of the environment and natural beauty were given equal standing with property values in the law.

Con Ed took the Appeals decision to the Supreme Court, which ruled against the utility by refusing to review the case. So the stage was set for new hearings before the FPC.

Meantime Con Ed revised the script. In the interests of natural beauty and to silence the "misinformed birdwatchers, nature fakers, land grabbers and militant adversaries of progress," the power plant would be put underground so as not to mar the mountain. Con Ed now proposed to hollow out Storm King Mountain, to create a huge underground chamber 700 feet long and 15 stories high, in which to house the turbines and transformers.

Could there be a more perfect example of Con Ed's sheer inability to understand the issues than that it would think such a proposal would satisfy people trying to stop the destruction of a mountain? Nor was the company alone. *The Wall Street Journal* thought the proposal a "signal victory of conservation," which would put an end to any further backtalk over trivialities from the conservationists.

But the cavern idea brought a new opponent into the ring. New York City's Catskill Aqueduct, bringing water from the mountains to the city, ran underground here, just 200 feet below Con Ed's proposed cavern. The aqueduct passed through an area of "strained rock," and the city engineers were distinctly unhappy at the prospect of anyone blasting out a cavern above their precarious water line. Con Ed, long-suffering, offered to move the aqueduct, but the offer was declined ungratefully.

At the new FPC hearings a geologist from Vassar testified that creation of an 8,000,000,000-gallon bowl that would be alternately filled and emptied could well reactivate a fault that ran under the Highlands, opening the prospect of earth tremors.

A major point made by the Scenic Hudson people (now allowed to testify) was that a clean mountain valley would be contaminated with billions of gallons of salty, polluted Hudson water, with disastrous results to the ecology of the valley. There was real danger that this water would leak through the ground into the underground water table and into nearby streams, with even more widespread pollution.

Con Ed cited the need for more power and blamed the 1965 blackout along the eastern seaboard on conservationists, who, it charged, had blocked plans to build new plants. Scenic Hudson countered with the Westfall Alternative, which proposed building a nuclear plant and five gas turbines, claimed to be less expensive than the Storm King project.

With public opinion running less than favorably toward Con Ed's revised plans, the utility began to have some doubts about the new hearings. And with unbelievable obtuseness it sent up a trial balloon for a new site—in the Palisades Interstate Park. Surely this will not be the last time an attempt is made to cut into a park, but if tried, that fight will be a long one.

In the Storm King battle the local citizenry were divided. Many supported Scenic Hudson; some Con Ed stockholders contributed their dividend checks. But many in Cornwall would have welcomed the plant as a means of broadening the tax base of the town. This kind of division is far from uncommon.

On August 19, 1970, the Federal Power Commission closed its second hearing on Storm King by again authorizing Consolidated Edison to build its pumped-storage plant. The decision was unanimous, and the commission said that the project offered "the best use of available resources to meet the requirements for electric energy with the minimum adverse impact on our environment."

The Scenic Hudson Preservation Conference said it was ready to continue the fight in the Federal courts and was promptly backed by the City of New York and Representative Richard Ottinger. At this writing another long-drawn-out court battle seems inevitable.

When an Army Engineers' dam in the Red River Gorge of Kentucky was stopped by a coalition of conservation groups, supported by the Louisville *Courier-Journal* and dramatized by a protest march led by Justice Douglas, the local people were furious. A photograph in *The New York Times* shows a picket with a sign reading *Sierra Club Go Home.* The attitude was, "Where do these damned outsiders get the nerve to come in here and stop a dam we've been trying to get since 1953, all for a worthless hollow not worth saving?"

To many local people the gorge was useless, but a dam contract meant jobs and supplies and lots of money spilling over into local cash registers. Not many cared about the Red River Gorge as Daniel Boone saw it when it was black with buffalo and noisy with flocks of wild turkeys. The buffalo and turkey are gone, but the trees have grown back now, and the magnificent natural stone arches, carved by water and wind, are still there, with foot trails that wind through acres of mountain laurel and rhododendron.

Progress versus beauty—but who defines progress? What has been called the last great hydroelectric dam in America poses the same problem in Idaho. The place: Hell's Canyon, on the Snake River.

From Yellowstone Park the Snake flows west to Oregon, then north. For 200 miles it forms the border between Idaho and Oregon and carves out Hell's Canyon, the deepest gorge on the American continent. Deeper than Grand Canyon, Hell's Canyon is similarly an ecological rarity. In places nearly 8,000 feet deep, it covers all the life zones of temperate America, from desert environment at the river level to Alpine conditions at the highest peak. An immense variety of wildlife makes it unique. The river is a spawning area for salmon and a last refuge of the legendary white sturgeon, a fish that may grow to a length of 10 feet and a weight of 200 pounds.

The canyon is also an archeologist's delight, being a meeting ground of two early cultures, the Great Basin and the Columbia Plateau. There are ancient petroglyphs on the walls and hundreds of intriguing archeological sites, of which only a few have been studied.

Holy Mother Snake, the Indians called it. They also gave the name Idaho to the area, "The light that comes down from the mountain."

The Snake and its tributaries already have something like thirty dams harnessing its waters. But the fantastic depth of Hell's Canyon and the power of the wild river—the Accursed Mad River it was once called—remain an itching temptation to the hydroelectric-power fans. The upper reaches of the Snake contain three dams: the Oxbow, the Brownlee and the Low Hell's Canyon dams. Now it was Hell's Canyon's turn.

In 1954 a group of utility companies formed the Pacific Northwest Power Company and applied for a permit to build two dams, the Low Mountain Sheep Dam, above the juncture of the Snake with the Imnaha River, and the Pleasant Valley Dam, 20 miles upstream. The FPC granted the permit, then withdrew it on the grounds that the Nez Perce site downstream had greater potential for the generation of electric power.

Pacific Northwest was willing, but there was a drawback. The Nez Perce site was below the junction of the Salmon and the Snake and a dam here would wipe out the thriving salmon industry on the Salmon River. So Pacific Northwest Power changed its application to the High Mountain Sheep site, above the juncture of the Snake with the Salmon.

Meantime a new combine, Washington Public Power Supply System, an agency of 18 utility districts in the state of Washington, had also filed for the Nez Perce site and now switched its own application to the High Mountain Sheep area.

At this point Secretary Udall and the Department of the Interior entered the picture, contending that private utilities should not build the dam at all, that it should be built by the Federal government, and that the site should be Appaloosa, still farther upstream from the High Mountain Sheep site.

With three contenders, the case went all the way to the Supreme Court, and in June 1967 a decision came down that broke new ground in the history of conservation battles. Justice Douglas, writing the court opinion, ordered the FPC to consider not only the case of which dam, but the case for no dam at all. The real test, he said, "is whether the project will be in the public interest," and tossed it back into the lap of the FPC.

The economic arguments as to the relative costs of producing electricity by water power or thermal power are already familiar. But on the question of no dam at all some of the comments are instructive. "It is Pacific Northwest Power Company's considered judgment that the middle Snake River should be developed and not left as an idle resource." To such interests, water that is not working is idle.

Senator Len Jordan of Idaho also had an interesting comment. He suggested the Nez Perce Dam site be reconsidered. Since the fish in the Snake are doomed anyway because of existing dams, he argued, why not go ahead and finish the job with the Nez Perce Dam?

Inevitably the opposition took form. The Hell's Canyon Preservation Council was formed by James Campbell, a nuclear physicist, with Boyd Norton, and was promptly backed by the Sierra Club and the Wilderness Society.

Currently, new hearings on the power companies' applications are scheduled before the FPC. But Secretary Walter J. Hickel of the Department of the Interior had announced that his department was calling a three-to-five-year moratorium on the government program for a dam at Appaloosa. And in Congress, Idaho's two senators, Frank Church and Len Jordan (who seems to have had second thoughts) have introduced a bill for a ten-year moratorium. Neither man seems interested in the conservationists' proposal to establish a 714,000-acre Hell's Canyon-Snake River National Monument to preserve the area permanently. So in a year, or five years or ten years, the battle will be fought all over again unless in the meantime nuclear power clearly demonstrates economic superiority.

If there is a feeling of unreality about these dam projects, let us take a quick look at the biggest boondoggle of all, the proposed

Ramparts Dam in Alaska. This ambitious job, estimated at $1,300,-000,000, would dam the Yukon River, create a lake bigger than Lake Erie, which would take thirty years to fill, and when filled would produce 34,000,000,000 kilowatts of electricity. This is enough electricity for a population of 6,000,000 people; Alaska's population is only 250,000 and isn't likely to reach 6,000,000 by the year 2000. As for selling the surplus electricity—Ramparts Dam would be 2,000 miles from the 48 continental states, which would require some long and expensive transmission lines.

The 400-mile-long lake created by the dam would be larger than the state of New Jersey; would drown 400 miles of the Yukon, 12,000 miles of tributaries and 36,000 lakes and ponds of the Yukon flats. It would wipe out the breeding grounds of millions of waterfowl, eliminate the salmon runs on the Yukon and destroy the range of moose, bear and caribou. "Nowhere," said a U.S. Fish & Wildlife Service report, "in the history of water development in North America, have the fish and wildlife losses anticipated to result from a single project been so overwhelming." *

But when Secretary Udall in June 1967 rejected the Army Engineers' proposal for Ramparts Dam, Senator Ernest Gruening of Alaska said he would continue to fight for the project. Mr. Udall, said the senator, was in the grip of extremists among "my fellow conservationists." And then, probably forfeiting his membership among any conservationists, he added that the Secretary "was more concerned about the alleged future of a duck than in the future of the people" in Alaska.

If Ramparts is the biggest, it is not the last on the drawing boards. The Eel in California, the Oakley in Illinois, the Tellico in Tennessee, the Big Walnut in Indiana, the Sun River in Montana and the Charlotte in New York are still projected.

The last-named is interesting because of the contrast with Kentucky's Red River Gorge. Whereas the local inhabitants in Kentucky felt cheated at losing the dam, the New Yorkers have rallied to fight it. Writing in the *American Agriculturist* (January 1970), Ray Christensen says: "Are larger dams necessary? Is the justification of the projects to go unquestioned? Are we to sit idle and accept the proposals when there are many questionable points?

"Since 1936 we have lived with the constant and continual harass-

* Report to Army Corps of Engineers, April 28, 1964.

ment from the Corps of Army Engineers who have been determined to build a large dam on our Charlotte Creek. Since then we have waged a successful battle each time we were faced with this threat."

One reason given for building the Charlotte Dam was low-flow augmentation. Translated, this means that during the low-water season in July and August additional water would be available from behind the dam to flush the sewage from the creek and "thus to create a gigantic cesspool out of the ocean." But, as Mr. Christensen says, "the dam is 120 miles from the city of Binghamton, the major polluter at this point. The Charlotte Creek contributes only four per cent of the water going past Binghamton." In any case, it is absurd to use clean watei to dilute stream pollution; the only satisfactory solution is to treat the sewage in an acceptable way.

Even the Potomac, our "national river," unbelievably polluted and the subject of a special plea from President Johnson, is caught up in the same dismal power struggle. President Johnson in 1965 asked Secretary Udall to develop a model program for the river. The Potomac, said Johnson, is truly the American river.

"I urge the Congress," he said, "to authorize the development of a uniquely historic area—the Potomac National River. Failure to act now will make us the shame of generations to come."

A Potomac National River Bill was introduced in Congress, and both the Department of the Interior and the Army Engineers proceeded to draft proposals for action. The Army dusted off a 1963 proposal calling for sixteen major dams on the Potomac and its tributaries, intended primarily for flushing out pollution.

Maryland residents visualized hundreds of acres flooded and homes abandoned because of this rash of dams and began to organize against it. They recommended instead a network of watershed structures like those suggested for the Charlotte to impound the headwaters of tributary streams and eliminate the big reservoirs planned by the Army. More beaver dams, in effect.

The Interior Department's recommendation, when it came, was a compromise. And a weak one, according to the National Parks Association, imminently concerned with the Potomac, "where completely new approaches were needed." Interior boggled at the Army's sixteen dams, but went along with three.

The aftermath was the usual proliferation of groups with varying programs and claims, and at last report the Potomac was running as polluted as ever.

In contrast, consider the Allagash, where a state and Federal partnership preserved a key wilderness stream. The Allagash, in Maine, is said to be the last piece of unspoiled wilderness in the East. Thoreau made the first plea for public ownership in the nineteenth century and only one hundred years later the voters of Maine approved a $1,500,000 bond issue, matched by an equal amount in Federal funds, to buy a strip of land 85 miles long and two miles wide, embracing the river.

In taking over the land, the state of Maine halted all cutting of timber in areas 400 to 800 feet on either side of the stream or the lakes that are part of the waterway. For another mile back of those boundaries lumbering must be approved by state authorities. Only canoes are permitted on the waterway, one motor not over 10 horsepower allowed; other motorboats are permitted only on three selected lakes of the system.

The state went beyond the original proposals of the Bureau of Outdoor Recreation by allowing snowmobiles in the forest, a decision they already regret. Cases are mounting of snowmobiles used to run down wild animals unable to escape because of deep snow.

The Allagash settlement is a landmark because it demonstrates the feasibility of state-Federal joint action in preserving wild country, without indication that the Federal government will devour state lands and state prerogatives.

The type of thinking that proposes a big dam for flood control overlooks the essential point that a dam does nothing for the watershed of the river. The world was touched by the devastating floods of 1966 in Italy that buried Florence in mud and ruined so much irreplaceable art. During the subsequent struggle to save as many art treasures as could be exhumed and cleaned no one seems to have said a word about the cause of the flood.

The simple fact is that Italy's topography is four-fifths hill or mountain country and only one-fifth plain. For hundreds of years those hills have been gradually denuded of their forest cover—the great sponge that absorbs the rain. Today the sharp spine of the Apennines is in many areas largely bare. The rain rushes down these steep slopes to surge like tidal breakers on the plain below, with the inevitable result of flood.

In many river basins of the United States we have a somewhat different problem. Every year, for example, the rivers in the Ohio and Mississippi basins rise over their banks and in spite of levees and sandbags and prayers and exhortations engulf towns and farms and sweep away houses, barns, livestock and cars. And each year the people come back and rebuild and hope it won't happen again. But it always does. The dam builders could take lessons from the beavers —small impoundments at the headwaters of streams are the safe and easy way to control the waters, with none of the disadvantages of the big dam.

Cities have grown up alongside rivers because rivers are natural transportation ways. Moreover, the flood plain that borders a river just asks to be built upon; it is usually level and accessible and the perfect place to tie up a boat.

But a flood plain is really no place to build. A flood plain is a sponge that soaks up water and returns it gradually to the river and the underground water table. However, as William Whyte points out in *The Last Landscape,** if you cover that plain with buildings you have a different story. The average building rooftop of 1,200 square feet will shed 750 gallons of water in a one-inch rainfall and absorb none of it.

Visualize a good-sized town of such rooftops and imagine each as a funnel pouring its load into the torrent racing through the storm sewers into the river without a chance for the ground to intercept and soak up any of it. The towns themselves contain all the ingredients for the very floods they fear. Dikes or levees that attempt to keep the river within its boundaries merely increase the problems of towns farther downstream.

Actually, there should be no building on flood plains at all, both to protect the underground water table and to avoid pollution of the river. But this is such a heretical idea that few people dare express it openly.

One who does is an imaginative ecologist named Ian McHarg, a big man with an aggressive brush of a mustache and a Scot's burr in his speech. He is head of the Department of Landscape Architecture and Regional Planning at the University of Pennsylvania. Says McHarg (*Audubon* Magazine, January–February 1966): "We

* Doubleday & Co., Inc., 1968.

cannot indulge the despoiler any longer. He must be identified for what he is—one who destroys the inheritance of living and unborn Americans."

Nature, says McHarg, provides free services to man, regulating floods with forests, providing soil for growing food, marshes for spawning fish and wildlife and controlling the water supply. But "marshes seem made to be filled, streams to be culverted, rivers to be dammed, farms subdivided, forests felled, flood plains occupied and wildlife eradicated."

This is not only wicked, says McHarg, it is stupid. Man's design should work with nature, not against it. Only by charting the water tables and the geological elements of the landscape can a plan be drawn that works with nature.

Grossly simplified, McHarg's principle is to leave marshes alone, leave prime farmland alone, leave steep slopes alone. "If you want to accelerate erosion and sediment yield, if you want to raise the level of flood plains, by all means build on steep slopes. . . . Any slope of more than 25 per cent should be prohibited to development, should not be cropped. It should be forest."

McHarg locates the aquifer, the place where surface water and ground water interchange. This must be protected against despoilment or pollution. Buildings and septic tanks here pollute the entire water supply. Where do you build? You build on the plateau, high above the plain, put the sewers or septic tanks there, leave the slopes alone, zone a strip at least 200 feet on either side of streams to avoid pollution.

Only 5 percent of the United States is urbanized now. We could double that without needing to fill in the marshes, without destroying open space. We build in clusters, with plenty of open space around each cluster. We make intelligent use of density rather than urban sprawl.

To the argument that open space is "wasted" McHarg retorts that open space is a major resource in itself, a means to check uncontrolled growth which obliterates valleys and flood plains and covers the landscape with spreading slums.

The whole concept of flood-plain zoning is revolutionary. It has been supported by court decision, but since few American towns worry much about the towns downstream, there has been little enforcement and much dike building and levee reinforcement.

Whyte says in *The Last Landscape:*

> Because of the vast amount of dam building that has been going on, the public is under the impression that the danger of floods is receding. But the opposite is true. By allowing developers to waterproof the flood plains, communities have been increasing the flood damage potential faster than the engineers can build dams to compensate. The public pays dearly, both in flood damage and in the cost of dams that otherwise would not have to be built. Just one shopping center and parking area built in the flood plain can create enough extra runoff to require the construction of anywhere from $500,000 to $1,000,000 worth of flood-control structures. The public pays the whole bill and retroactively provides a subsidy to developers for building where they shouldn't.

There is some reason to believe that the era of the big dam is about over, although some engineers are still unaware of it, and more dams are on the drawing boards.

As a means of flood control the big dam is ineffective. It alters the natural drainage pattern, upsets the water tables, loses water by evaporation and thus adds to the salinity of the water—and does nothing about reforestation, which is the heart of flood control. As Justice William Douglas says in *A Wilderness Bill of Rights:* * "We are entering an era where we need not destroy free-flowing streams and lovely valleys to have the energy we need. With desaltation now in practical grasp we may not even need dams for irrigation."

Surely we do not need them for power; hydroelectric power is obviously being eclipsed by nuclear power, although nuclear power plants have certain pollution problems to solve. Succeeding generations may look at our dams the way we look at the great pyramids of Egypt, wondering what particular brand of folly or pride made us build them.

* Little, Brown, 1965.

CHAPTER

2

Water Pollution and the Great Water Shortage

THE rain came down, day after day, all though the summer of 1969. River basins were flooded and towns inundated by streams rising high over their banks. With ample rainfall continuing into the spring of 1970 and reservoirs filled, most people would have found it hard to believe there is a water shortage.

Actually, there is the same amount of water there has always been. There is simply less water available because so much of it is so incredibly polluted. In any particular area there may be wet and dry cycles, with temporary water shortages during the dry periods, but over the long term there would be enough water to support us if it were available. Pollution is diminishing the supply while the demand continues to rise with rising population and the needs of industry.

Consumer uses take less than 10 percent of the water supply. This includes baths, dishwashers and lawn sprinklers. The big users of water are industrial plants and agricultural irrigators.

Consumption of water by industry is enormous. It requires 65,000 to 70,000 gallons of water to produce one ton of steel; 140,000 gallons to process one ton of wool; 600,000 gallons to manufacture one ton of synthetic rubber.

While this water goes back into the supply after it leaves the mill or plant, it goes back carrying a heavy load of chemicals and other wastes. The steel mill may discharge 300,000,000 gallons of water a day, much of which will contain phenol, cyanide, ammonia and oil. Phosphates from the detergent industry, acids from paper mills, pesticides washed down from farms and raw sewage from thousands of towns all make an unsavory combination that is toxic in itself, but that also reduces the oxygen content of the water so that it will not even support the marine life that might resist the poisoning.

Said *Time* Magazine (October 1, 1965):

> Industry now pours at least twice as much organic material into U.S. streams as the sewage of all the municipalities combined. Americans who once could be excused a superior attitude about sanitation after traveling abroad now come home to find that their own drinking water may come from rivers into which steel mills pour pickling liquors, paper mills disgorge wood fibers that decay and use up oxygen, and slaughterhouses dump blood, fat and stomach contents of animals. Pollution has become such a problem that it is all but impossible to calculate the probable cost of cleaning up the streams. A conservative estimate: at least $40 billion over the next decade.

In the few years since those words were written hardly a start has been made, and any theoretical timetable will have to be set back again and again.

The mechanics by which a stream is murdered are well known. Clear water has a rich oxygen supply. It sustains a variety of plant life and animals such as fish, shrimp, insects, snails and frogs.

As sewage and other pollutants are fed into the water, decomposing bacteria multiply enormously, consuming the sewage but using up the oxygen in the process. The animal and plant life dies off and is replaced temporarily by sludge and blood worms, mosquitoes and leeches. A heavy algae growth, black and odorous, covers the bottom. As the process goes on, all the oxygen is eventually consumed. The water becomes completely foul, with gases bubbling to the surface and only anerobic bacteria surviving.

By 1980, said the National Research Council, sewage and other water-transported wastes will have reached a volume sufficient to consume, in dry weather, all the oxygen in all the 22 river systems in the United States. Which can be interpreted in no other terms than the death of our waterways.

This situation has already been reached in many streams and lakes. A local comment in Cleveland is that the Cuyahoga River is so polluted that it is the only body of water in the world that is a fire hazard. It actually caught fire and burned in the summer of 1969.

The extent of pollution is so enormous that it is difficult to grasp. The Great Lakes contain 20 percent of all the fresh water on the face of the earth. To pollute so vast an amount of water seems like a difficult job, but it has been done.

A few years ago, in a special issue (December 14, 1965), *Look* Magazine said a requiem for Lake Erie: "A 2,600 square mile patch in the middle of the lake, equal to one-quarter of its area, has become devoid of oxygen for as much as 10 feet up from the bottom. Whitefish and pike, which were the basis of a multi-million-dollar fishing industry, have been exterminated. Swatches of algae drift ashore onto beaches of Michigan, Ohio, Pennsylvania and New York, and rot, making swimming a nauseating pastime."

Since then, of course, the blight has spread to more of the lake. The Detroit River is also at this end point, harboring only sludge and bloodworms, which require little oxygen, and fingernail clams. The Buffalo River in New York is in a similar condition—slime taken from the bottom harbors no life.

At Niagara Falls, tourists taking the famed under-the-falls ride on the little steamer *Maid of the Mist* sail through a mist of pollution coming over the brink from sewage outlets above the Falls.

The office of Kansas' Governor Robert B. Docking reports that it is unsafe to drink from or swim in any stream or river in the state. From 1963 to 1966 Kansas had 93 fish kills in its rivers and streams.

Lake Tahoe, on the California-Nevada border, has lost its reputation as one of the three purest lakes in the world. It is discolored now by wastes from the mushrooming cabin colonies along the shore.

The three most beautiful streams in the world, the Connecticut, the Hudson and the Rhine, are perhaps the most polluted and befouled.

The Connecticut crosses all of New England—400 miles from the Canadian border to Long Island Sound. Salmon were once so plentiful in the river that until the year 1800 there was a law against feeding them more than three times a week to servants. It was said that in the spawning season a man on snowshoes could cross the river on their backs.

Symptomatic, perhaps, of our technological muddle is the fact that there are at least four government studies of the Connecticut River under way. There is an Army Engineers' study to determine the feasibility of dredging a boat channel from Hartford to Holyoke; there is an Interior Department study sponsored by Senator Abraham Ribicoff to see if the river can be made a national recreation area; there is a joint Federal-state Water Resources Inventory, which will run until 1973; and there is a $5,000,000 Connecticut River Basin study.

A new Parkinson's Law, the Prohibitive Procrastinator, defines a "study" as a way of avoiding doing something. True, Connecticut has a Clean Water Law that makes polluting the stream illegal. But enforcement is something else again. The state's total of human and industrial sewage is 200,000,000 gallons a day, which is too much for all of its 14 river systems to handle. The result is that communities are still taking their drinking water from streams that are grossly polluted.

Nor is the Connecticut the worst example. The water from the Ohio River is used and reused three times before it reaches the Mississippi, and after that—who knows? The Connecticut Clean Water Law is an advance, but no one expects it to eliminate pollution in the next decade.

The inadvisability of swimming in the river would come as no new thing to New Yorkers, whose Hudson has been a hazard for many years. The decline of boating around New York is attributable to a number of factors, but one of them is the fact that if you fall overboard you'd better hurry and get your typhoid shots. However, the city, which recoiled from the thought of drinking polluted Hudson water, has little reason to complain of upstream towns. The city discharges 400,000,000 gallons *a day* of raw sewage, untreated in any manner, into the river and its bays.

However, the Hudson is certainly not clean as it reaches the city. A hundred and forty miles upstream, at Troy, it is already heavily contaminated. For the next 50 miles the contamination continues, to the stretch near Kingston, 90 miles from New York. From Kingston to Storm King Mountain, past the city of Poughkeepsie, little more is added—although if Con Edison should finally win its fight to hollow out Storm King Mountain for a pumped-storage plant, a new form of pollution will be added. Sewage and pollution begin again at Peekskill and increase through Westchester on the east with Rockland County and then New Jersey on the west, until the river receives its final discouragement from New York itself.

The troubles that afflict the Hudson are so many that the situation on the Connecticut is simple by comparison. Most of the river towns that line the banks of the Hudson are offensive slums. The shoreline for miles is littered with broken piers, junked cars, discarded tires, used lumber, empty oil drums, debris and trash of every kind. Shoreline rocks and pilings are coated with oil and tar. An iridescent film of oil layers the surface and offends the nostrils.

The pollution begins at Corinth, about 60 miles above Albany, where the paper mills assault both air and water. The river bottom here is covered with a mat of felted fibers, a papery material that comes from the waste discharge of the mills. This mat has blotted out all bottom life. As bacteria attack it and it decomposes, great bubbles of gas erupt to the surface, bringing up loose chunks of the felt mat, which float off on the surface like an invading fleet.

The decomposition of this material robs the water of oxygen and destroys whatever fish and marine life have escaped the other kind of pollution from the mills, the "black liquor," which is a mixture of acids and wood-pulp residues.

The New York State Department of Health has counted 435 polluters in the 150-mile stretch of river between Troy and New York. These include not only factories but towns, and institutions like Sing Sing prison, and the military academy at West Point, which is still waiting for Federal funds to build a new treatment plant to handle its daily load of 1,000,000 gallons of sewage. Altogether, 2,000,000,000 gallons of waste are poured into the Hudson daily, most of it untreated and carrying acids, dyes, chemicals, oil, soap, detergents and solid wastes.

In the last few miles of its journey to the sea the Hudson receives its final baptism of raw sewage from New York and the towns across it on the Jersey shore. To walk the shoreline in New York's Riverside Park and smell the river is an educational experience.

In 1965 New York voters approved a $1,000,000,000 bond issue to clean up the Hudson. The Federal government promised help in subsidizing municipal sewage plants. A Pure Waters Division was established in the State Health Department and ordered to make life difficult for polluters. Governor Rockefeller estimated that the Hudson would be clean by 1972.

They were brave words. A billion dollars isn't enough to clean up a major stream so grossly polluted that it serves as a horrid example to worried conservationists, who question seriously whether any amount of money or work can bring back a river whose ecology has been so brutally assaulted over so long a time.

Even now not much of a start has been made. Enforcement of antipollution laws is feeble or nonexistent. Paper mills threaten to close rather than undergo the expense of confining and treating their wastes. The Federal contribution for sewage plants, expected to be 40 percent of the state contribution, has so far run closer to 7 percent. The most optimistic guesses now are that it will be at least 1975 or 1976 before the waste-treatment plants are built—if then.

A new threat to river water is growing in scope—thermal pollution, the raising of river temperatures by large amounts of hot water from nuclear plants.

In 1968 conventional steam condensers in electric generating plants used some 75 percent of the 60,000 billion gallons of water employed in the United States for industrial cooling. While this is a lot of water, it poses no major threat except for particular local situations. The real threat comes from the projected building of nuclear generating plants. These generators waste 60 percent more energy than plants using fossil fuels, and this energy must be dissipated as heat.

It is quite possible, but relatively more expensive, to dissipate this heat into the atmosphere by means of cooling towers, with as yet uncalculated effects on the atmosphere. It is considerably simpler and cheaper to discharge the hot water directly back into the river.

John R. Clark, assistant director of the Sandy Hook Marine Laboratory, U.S. Bureau of Sport Fisheries and Wildlife, estimates that within 30 years we will be producing nearly two million megawatts of electricity, with the problem of dissipating about 20 million billion BTUs of heat a day. If this were done through existing waterways, it would require a third of the average daily runoff in the United States.

Many rivers in the United States normally reach a temperature of 90 degrees in the summer. A 1,000-megawatt power plant will easily raise the temperature of a river flowing 3,000 cubic feet a second by 10 degrees. Build several such plants on a river, says Mr.

Clark, and it is obvious that marine life is doomed. For example, the lethal temperature for crayfish is 95 degrees, for lake trout, 77 degrees. Carp can exist in water carrying oxygen at a concentration of half a milligram per liter of water if the water temperature is as low as 33 degrees. Raise the water temperature to 95 degrees and the carp need three times as much oxygen. Other fish can get by on one or two milligrams of oxygen at 39 degrees, but if the temperature is raised to 65 degrees only they need about four milligrams to survive and five milligrams to function normally. The Oregon Fish Commission has said that a rise of only five and a half degrees in the Columbia River would end reproduction of the Chinook salmon.

The killing of fish from pesticides became almost commonplace in the Mississippi River and even in the Rhine during the summer of 1969, but fish kills by thermal pollution is a new phenomenon. In the summer of 1968 a considerable number of menhaden were killed in the effluent water of a new power plant on the Cape Cod Canal. The menhaden, used to temperatures in the 80s, were trapped in water that reached 95 degrees.

New York's public energy utility—Consolidated Edison—built its first atomic power-generating plant at Indian Point on the Hudson. Four more are proposed in a one-mile stretch of shoreline north of Haverstraw Bay. The proposal has alarmed conservationists, who are concerned that the grouping of five nuclear plants too closely together will raise the temperature of the river critically. The Indian Point plant, using 300,000 gallons of water *a minute* to cool its reactors, returns it to the river at least 15 degrees warmer. Multiply this plant by five, and so much water would be warmed that cold-water fish would die and algae reproduce uncontrollably. Algae, upon dying, use up oxygen in decomposition, killing more fish and contributing to the rapid eutrophication, or aging, of the river.

Con Edison insists that its studies indicate there is no danger of this happening, but Con Ed studies have been questioned before. In 1963 the Indian Point plant was responsible for a massive kill of striped bass. The fish were attracted by the warm water and were then trapped in the intake conduits, where they died of fatigue and stress. Tons of the fish were carted away in trucks in a clumsy and unsuccessful attempt to conceal the accident.

The company then announced that it had solved the problem with new screens to keep the fish from being drawn into the conduits.

Conservationists contended that no screens would keep out eggs and larval-sized fish.

On February 3, 1970, *The New York Times* revealed that another fish kill had been under way since January and that Con Edison had had to shut down the plant for three days to bring it to a halt. The company estimated the kill at about 150,000 fish. Former Representative Richard Ottinger of New York put the number at "hundreds of thousands to millions."

Next day's *Times* carried a new Con Ed assurance that this time it had really solved the fish-kill problem with better screens and a relocation of its discharge conduits farther from the plant.

At Forked River, New Jersey, the State Department of Conservation lost one round with the Jersey Central Power and Light Company in its contention that a new nuclear power plant would constitute an environmental hazard if it raised the temperature of the waters of Barnegat Bay higher than 86 degrees.

The State Board of Public Utilities Commissioners ruled that Jersey Central would be permitted temperatures of 95 degrees until this should be proved too high. A water-tower cooling system proposed by the Conservation Department was rejected by Jersey Central on the basis of cost—about $7 million. Jersey Central agreed to contribute $75,000 to a three-year study by the Conservation Department to determine the effects of the rise in temperature on marine life in the bay.

In addition to the direct effect of higher temperatures on fish, warmer water stimulates the growth of plankton and bacteria, which use up oxygen, choke waterways and frequently poison fish and shellfish. The river then goes into the downward spiral already described.

As far back as 1965 President Johnson, who, however hesitantly, at least did launch some major conservation programs, said, "Every major river system is now polluted."

The oceans of the world are huge, but even these are already sadly polluted. Twice a day from New York a tanker belonging to the Titanium Lead Company rides out to a spot about seven miles east of Ambrose Light and dumps the acids from the plant operations. Sewage sludge from New York is dumped at another point about three miles southeast of Ambrose. Mapping the growing mound of sludge, the Fisheries Bureau has found an area of 20 square miles in the Atlantic Ocean in which nothing is alive—a dead area like that in Lake Erie.

This is no longer an abstract matter, of interest only to conservationists. For man's survival he needs the water from rivers and lakes and oceans, and the idea is beginning to sink in, however belatedly, that no one has the *right* to dump wastes into the waters that belong to all of us.

During New York's drought of 1965–1966 there was some panic that the rains had indeed gone away never to return. *The New York Times* of March 23, 1966, carried a deadpan report that the city contemplated a go at rainmaking. An aide of Mayor John Lindsay announced that he would be interviewing meteorologists and experts in cloud-seeding to see if he should recommend this to the mayor.

"I do not know if we will attempt rainmaking," he told the *Times*. "But I am looking into it. My study is simply one more example of the determination of the Lindsay Administration not to overlook any step that might help the water situation."

Growing metropolitan areas compound the water problem, because large cities are now competing with each other for water supplies, reservoir sites, distribution channels and sewage facilities. In spite of the complexity of the modern city, the techniques of water supply are hardly more advanced than last century's backyard well and pump. As the metropolis expands outward, such local sources for water as do exist are quickly overrun, and it becomes necessary to reach across county or state lines, as California and New York are already doing—which may create state and Federal problems.

Throughout New York's dry summer, as the water in the reservoirs dwindled, voluntary and some not-so-voluntary rationing was urged upon a confused citizenry, whose newspapers almost daily printed pathetic pictures of the cracked mud bottoms of their once boundless reservoirs. The humorists had their day: "Save water—shower with a friend."

The sudden uproar was all the more bewildering to New Yorkers since they had always prided themselves upon one of the best and purest water supplies in the world. The claim had point, for New York's water came in large part from sources high in the Catskills and, even after its 150-mile journey to the city, arrived fresh and cold and pure. Additional supplies from the Delaware River were also relatively pure.

Again, there was no shortage of water, but only of clean water. Past Manhattan Island the once lordly Hudson moved at the rate of 15 million gallons a second, but so thoroughly polluted that to use it without heroic cleaning measures was out of the question. The smell of it was something that hitherto had annoyed only conservationists, but of late the drinking water from New York's taps had begun to take on an odor familiar to residents.

Fifteen years earlier there had been a briefer water-shortage scare. A pumping station had been constructed at Chelsea, 60 miles upstream from Manhattan, at a cost of $2,600,000. It was never used. The emergency faded and the station was dismantled.

Its brief existence was recalled in that summer of 1965 and it was hurriedly rebuilt at a cost of $3,000,000, but too late to get into operation for the dry spell. In fact, it did not go into action until the following spring, when it ran for only eight days before being shut down. In those eight days it pumped 384,000,000 gallons into the system, filling the West Branch Reservoir in Putnam County to its brim.

To pump any more would have been only to have it spill over the dam, and a spokesman for the Department of Water Supply was quoted in *The New York Times:* "There's no point in loading up from the Hudson when good clear water from the Delaware watershed may be running to waste."

His words take on enhanced meaning when one considers the treatment necessary to make Hudson River water drinkable.

The water was first drawn through a fine screen to remove the larger particles of foreign matter floating in it. It was then given a heavier-than-usual dose of chlorine, plus alum, which coagulates small particles and helps to settle them. It was then mixed with Delaware River water in a five-to-one ratio. The combined water was then sent through an aqueduct to the West Branch Reservoir, where it was held at least 20 days. It was then given a second dose of chlorine. Every two hours five chemists and five bacteriologists took samples for testing.

Diluted again with other water (presumably Delaware or Catskill), the Hudson's contribution was then moved down the Croton system to the Kensico Reservoir. Here it stayed another month, being further mixed with other water and chlorinated for the third time. It then moved on to the Hill View Reservoir in Westchester County, where it was chlorinated for the fourth time. Finally, 30 to 40 days later, some Hudson River water and a lot of chlorine reached the city's taps.

Meantime, in that summer of 1965, the inhabitants of New York were exhorted to save water. The irony of the situation was that all the pleas to skip baths and car washing, to stop sprinkling lawns and turn off leaky faucets, and the tracking down of water leaks by political candidates (the campaign for mayor was in full swing) were little more than window dressing.

In the same rush of enthusiasm that prompted the billion-dollar bond issue to clean up the river, Governor Nelson Rockefeller unveiled another plan to restore the Hudson Valley to its original beauty. It called for the creation of a statewide system of historic parks and landmarks, with a network of roads to cultural, historic, scenic and architectural sites, to be called the Hudson River Tourway; the acquisition of 100 endangered scenic areas along the river, including Storm King Mountain; the cleaning up of roadside billboards and junkyards; a new sewage plant in Manhattan; a system of trails for walkers near urban areas and the use of the undersides of river bridges to carry utility lines across the Hudson.

If any work was ever done on this plan, it too has escaped public notice.

In the newer Midwest, rivers are no cleaner than the Hudson, as we noted in Kansas. By 1966 the situation was considered so desperate that four states—Nebraska, Missouri, Iowa and Kansas—joined in a program to reduce the amount of offal being dumped into the Missouri River by one of the country's largest concentrations of meat-packing plants.

Twenty-four slaughterhouses in Omaha handled about 5,000 head of cattle a day, producing about 122,000 pounds of waste grease. There was an additional 200 tons a day of wastes from the stomach contents of the animals. Much of this was dumped directly into the river because it would have overloaded the sewage-treating facilities. Ironically, it left a new $13 million sewage plant with not enough to do, operating at only a fraction of capacity.

A city official of St. Joseph, Missouri, 100 miles downstream, brought a jar of grease pellets to a Federal-state meeting in Omaha to show how Omaha's slaughterhouse wastes were clogging the intakes of St. Joseph's water system.

The result of the meetings was an agreement calling for a reduction to 13,000 pounds a day of waste discharge into the river. The total of other solid wastes was to be limited to 160 parts per million of water. A projected new treatment plant, if approved by the voters, was expected to cut the 122,000 pounds of daily grease to 25,000 pounds. Processed by the regular sewage plant, this would then reduce it further to 8,000 pounds a day. It is presumed that the citizens of St. Joseph were duly gratified that they would be coping with a maximum of only 13,000 pounds a day of offal in their drinking water.

Officials in Denver found the South Platte River and its tributaries along a 24,300-square-mile basin in such bad condition that they were admitted to be a possible source of epidemic disease. Pollution came from 14 municipalities and 7 sanitary districts—from industrial plants, cattle feed lots, 13 slaughterhouses and 10 sugar beet mills.

Near the Denver suburb of Brighton, bacteria in the river were measured at a density of more than 3 million per 100 milliliters—considered 3,000 times the safe level for swimming. A Federal report said that the pollution of the river had created ideal conditions for the breeding of mosquitoes, flies and rats, all of which can transmit plague, encephalitis and enteric disorders.

The reaction? The sugar-beet industry complained that the facilities needed by their mills alone would run to $30 million.

Industry is coming to realize, however reluctantly, that it must curb its own pollution, although its initial reaction was one of panic. The *Journal of Commerce* (September 1969) quoted a report from the American Chemical Society that "the tab for cleaning the environment will have to be paid by the U.S. citizen, and his reluctance to pay may be one of the most powerful restraints in the near future."

The question of who pays for cleaning up the rivers and lakes is much less complicated in Europe. In West Germany the Ruhr River flows through one of the most industrialized regions in the world, one containing many steel mills, yet the river is clean enough for swimming anywhere. A very simple rule is in effect: Whoever pollutes the river must pay to purify it.

Steel mills in the Ruhr employ water-circulation systems that use the same water over and over again. They draw only 2.6 cubic yards of water per ton of steel, compared with 130 cubic yards formerly used. This simple, logical policy has yet to be adopted by American industry.

Against this dreary recital of abuses, what is being done? In 1965 the Johnson Administration set up the Federal Water Pollution Control Administration as part of the Department of Health, Education and Welfare. The new administration announced that 12 regional laboratories would be constructed to study water-pollution problems. A national program would also be set in motion to provide grants to communities for building sewage-treatment facilities, provide technical assistance to states for water-pollution-control programs, enforce Federal pollution-control laws, develop long-range programs for the nation's major river basins, provide grants for research on sewage-treatment problems, establish water-quality standards for interstate streams and help train personnel in the field of water-pollution control.

The new agency barely had time to issue by the following spring a "Statistical Summary of Municipal Water Facilities in the United States," based on 1963 data, before it was embroiled in departmental politics. It was shifted to the Department of the Interior, where most people thought it belonged (rather than at HEW), and, with the advent of the Nixon Administration, came under Secretary Hickel.

In April 1969 Mr. Hickel and Lee A. DuBridge, then science adviser to the President, announced the formation of an Environmental Quality Council and defined as national policy the rights of Americans to have clean air, water and other ingredients of a healthy environment. Yet three months later, in July, Karl L. Klein, Assistant Secretary of the Interior, told a water-pollution conference that pollution was a local matter, because the Federal government could not afford it.

In September Dr. DuBridge repeated his pledge and outlined a war on pollutants. The same month Secretary Hickel said he would ask the Department of Justice to seek court injunctions against violators who failed to desist from polluting within 180 days of warning.

January 1, 1970, President Nixon promised "a now or never" fight against pollution and created a three-member Council on Environmental Quality under the chairmanship of Russell E. Train, formerly of the Conservation Foundation and, very briefly, Undersecretary of the Interior. It was widely speculated that conservationist Train had been appointed Undersecretary of the Interior to placate conservationists, who had been apprehensive about the choice of Hickel on the basis of his record as governor of Alaska.

The Council on Environmental Quality published its first report in August 1970. It was, said Gladwin Hill in *The New York Times* (August 16, 1970), "less a record of accomplishment than a laundry list of problems to be grappled with. Most of them were ones with which the average citizen is only too familiar—air pollution, water pollution, solid waste, pesticides and the other bugbears gnawing away at the 'quality of life.' "

Because the report was aimed at the President, at Congress and "the great army of slow-moving Federal, state and local officials and bureaucrats whose sluggishness has contributed to the environmental mess," said Hill, "the council was impelled to walk softly and carry a cheerleader's megaphone." In short, it could have been a stronger report.

Meantime another bureau, the Environmental Control Administration, which had remained within HEW, had been taking surveys of the nation's drinking water. While ECA apparently did not intend to publicize its findings, the news was broken by Virginia Knauer, Presidential assistant for consumer affairs, that 30 percent of community water samples taken showed excessive quantities of bacteria and chemicals. Moreover, she said, nearly all the samples showed traces of pesticides, and some contained arsenic in excess of Federal standards.

A spokesman for ECA confirmed Mrs. Knauer's report and admitted there was cause for concern, though not alarm, in certain areas of the country. Arsenic is not eliminated from the human body; it is cumulative and it can cause cancer of the skin or liver. However, the ECA spokesman said reassuringly, nearly all the arsenic found in drinking water came there naturally, from rock and soil through which the water flowed. He admitted being surprised at the prevalence of pesticides in drinking water and considered this a threat, since municipal purification systems are not geared to eliminate this kind of chemical impurity.

To round out 1970's parade of paper tigers, the administration created, in June, a superagency called the Environmental Protection Administration. The new agency brought under one roof water pollution from Interior, air pollution and solid-waste disposal from HEW, pesticide standards from the Food and Drug Administration, pesticide control from Agriculture and radiation regulation from the Atomic Energy Commission.

The Washington *Post* quotes a White House source as saying, "The big reason for the reorganization is efficiency. We weren't getting anywhere against pollution under the old setup."

Biggest loser in the reorganization is Interior, which lost water pollution, commercial fisheries, salmon fishing, marine mining, and any hope of running the environmental program for the nation.

Reclaiming our once pure water supplies is one aspect of the problem. Another is bringing water to places where it is naturally scarce, or creating new local supplies.

In Chanute, Kansas, the city government had no additional water sources of any kind to draw upon in response to growing demand. So they put the city sewage through a purification process and piped it back into the reservoir. Chemical analysis, said the city fathers, showed this water to be purer than it had been when it came out of the river originally. At least no one, it is said, got sick in Chanute.

The reclaiming of water is attracting interest elsewhere. In Los Angeles County, 12,000,000 gallons of purified effluent from sewage plants are put back into the ground every day. Sanitation officials claim this maintains the water table and provides a barrier against salt-water intrusion. In Las Vegas, in parts of Texas and southern California, lawns and golf courses are watered regularly with processed sewage water.

Another approach is through desalinization of sea water, a science still in its infancy but attracting attention throughout the world. Of an estimated 1,101,117,143,000 gallons of water on earth, 97.2 percent is in the oceans and another 2 percent frozen in glaciers and ice caps at the poles. That leaves only .8 percent of fresh water available for immediate use. It would therefore seem desirable to tap the oceans, if only an economical method of desalting the water could be found.

In hot, sunny countries such as Arabia and Israel, where sunshine is a dependable year-round commodity, distillation by sun heat is relatively simple, if slow. A newer method is freezing. Alexander Zarchin, an Israeli engineer, discovered that when sea water is frozen, the ice crystals form separately from the brine. Skimmed off, they will then melt down into fresh water. This process appears to be practical enough to warrant further development.

Most of the world's 200 or so desalting plants operate by a flash distillation process. This calls for heating the sea water and spraying it into a low-pressure chamber. The drop in pressure creates a lower boiling point, and the water flashes instantly into steam. It takes three and a half gallons of sea water to make one gallon of fresh distilled water. Costs, originally about $5 per thousand gallons, have been reduced to about one dollar. The goal is 35 cents or less.

Desalting plants are increasing, for there are many otherwise desirable areas in the world where no source of water but rain and the sea exist. For example, St. Thomas, in the Virgin Islands, has in the past brought water by barge from Puerto Rico at a cost of $2 a thousand gallons. It is estimated that a new desalting plant will bring this cost down to at least 90 cents.

Westinghouse, leading the field in the design and construction of flash distillation plants, has already built and installed more than fifty in various places. A nuclear-powered plant is going up on Long Island, built by American Machine & Foundry.

Another nuclear-powered plant is scheduled for Los Angeles, with a projected capacity of 150,000,000 gallons a day. And at the government laboratories in Oak Ridge, Tennessee, advanced methods of desalinization are being sought.

Two promising new developments in this field are reverse osmosis and ultrafiltration—both techniques for separating relatively pure water from pollutants that may occur in particles too small to be caught by ordinary filters. Ordinary filtration will remove particles no smaller than one micron in size. Ultrafiltration extends this range down to large molecules. And reverse osmosis, using thin membranes, will separate fluids from small molecules and even inorganic ions.

Membranes are completely different from ordinary filters. A filter offers openings through which particles may pass if they are small enough. The finer the filter, the smaller the particles that can pass, up to the limit of about one micron. A membrane, however, may not show pores even under the microscope. For this reason it will pass fluids while holding back small organic molecules—sodium, calcium, antibiotics, yeast, bacteria, viruses or enzymes. On the other hand, a membrane may pass some large molecules while blocking smaller ones. The action is comparable to solubility in a liquid; some molecules may react to the membrane as though they were soluble in it, while smaller ones may be insoluble. In handling sewage, membrane plants can turn out water with no detectable suspended solids, while conventional processes show 15 to 35 milligrams of solids per liter of water.

The first reverse-osmosis desalting plant was launched in August 1969 by Du Pont at Plains, a small town in Texas. The business end of this plant consists of 16 vertical cylinders, each containing a bundle of fine nylon tubes with some 85,000 square feet of semipermeable membrane surface. Output is said to be 100,000 gallons a day of drinking-quality water, with an eventual cost of about 50 cents a thousand gallons. Residents of Plains had been buying drinking water in jugs at 30 cents a gallon.

Dow Chemical's plant in Walnut Creek, California, will produce tubes made of cellulose acetate, and American Standard has a plant at Hightstown, New Jersey, for the manufacture of tubes made of glass fiber and plastic.

Dow research is aided by a grant from the Interior Department's Office of Saline Water. A new sewage-treatment plant combining activated sludge and ultrafiltration is being installed by Dorr-Oliver on Pikes Peak to handle effluent from the tourist tide up that mountain. This project is partly financed by the Federal Water Pollution Control Administration.

Reverse osmosis may be the most important process yet developed. On paper the energy required to desalt sea water by using membranes is 3.5 kilowatts for 1,000 gallons, as opposed to 100 kilowatts for distillation.

That man will turn to the sea for water seems inevitable. But this will not change the practical necessity for cleaning up our rivers and restoring the forests, which constitute their essential watersheds, and in so doing, preserve the natural beauty and the natural wildlife that are part of the natural balance.

CHAPTER

3

The Everglades

IF you drive south from Miami along Route 1 and turn west at Homestead into Route 27 you will come presently to the Park Department's chastely simple sign in the piney woods and you will know that you are about to enter Everglades National Park.

On leaving Miami, you had passed through developing suburbs, and then, as the country opened, past truck farms and other signs of agricultural activity—trailer trucks, packing houses and loading platforms along the railroad right-of-way.

Now, as the sign drops behind, you cross an invisible barrier and are in a world purged of commercialism. You have also crossed a war zone. The Everglades, a subtropical wildlife haven so prolific it must be seen to be believed, is losing that war. Its adversaries are the two you just passed: suburbia and agriculture.

But for the moment the image seems less than hopeless. The Miami jetport has been defeated, the rains have come, the immense swampland is filled with life-giving water, the deer have been saved and a ban on alligator shoes and bags has granted a reprieve to the vanishing saurian. So for a little while the prospect pleases, and if the Park and the Army Engineers still argue over the water promised the Everglades, here and now it doesn't show as you drive southward.

Some advance briefing is needed to appreciate the Everglades.

If you expected gloom-shrouded jungle, as many do, you will be surprised. You can drive the 39 miles from the entrance station to Flamingo and see brilliant sky, a vast expanse of grass and the glint of water.

It is a sea of grass, wider than the horizons on all sides. Actually, it isn't grass. It's a sedge, *Cladium jamaicensis,* a great sword blade with a deep center crease and wicked little teeth along both edges. It stands with its feet in the water and reaches for the sun, and it grows as tall as 15 feet where the water is deep. "Saw grass" is its common name, and the Everglades contain the largest concentration of it in the world, millions and millions of acres. The park measures about 1,400,000 acres, and it is only 7 percent of the entire Everglades.

You sail over the saw grass on Route 27 as though on a boat, mile after mile. Blue and white—the sky is an intense blue, the sunlight is the incredible white light of the subtropics. On the horizon, cumulus clouds build their piled mountains of snow fields. Islands of trees swim past—sometimes the road goes through them.

These are the "hammocks" of the Everglades, and they take their designations from the kinds of trees in each. There are mahogany, sweet bay, cypress, willow, pond apple, or wax myrtle hammocks. There are gumbo-limbo and paradise trees. The hammocks are carpeted with moss and festooned with orchids and bromeliads—air plants—or sometimes spiny with rare water-loving cactus. There are coffee trees and wild-grape vines and lotus and poppies, even blackberries and raspberries. And there are birds and raccoons and black bears and deer and a Florida panther that is a paler cougar than those of the North. There are thousands of hammocks, and some of them have never been reached or explored by man.

At the side of the road the Park Department's markers spell out the elevation, and it is intriguing to see it measured in inches above sea level instead of feet. Presently these inches will mean more to you in understanding the death struggle of the park.

Closer to Flamingo and the end of the world at Cape Sable, the mangroves appear, and you begin to see birds in the ponds and sloughs. These are not the tiny birds of the northern forests, shy and hard to see. These birds are big and tame. Herons and egrets stand man-high, wrapped in white plumes like robes. The anhinga dries its wings on a low branch in the sun. In the shallows the roseate spoonbill probes the mud for food, quite casual about the incredible pink of its wings and indifferent to the ecstasy of impressionable tourists.

A rare osprey goes by and a brown pelican abruptly folds its wings and drops like a stone into the water. It emerges without a fish, but with its dignity unimpaired, and takes off again, managing to look somehow like a prosperous middle-class citizen with important matters on his mind.

A raccoon idles at the side of the road, waiting for a handout from the tourists. An alligator sleeps in the sun. Everywhere water flashes and sparkles in the intense light, for water is the essence of the Everglades.

Pa-hay-okee, the Indians called the Glades—"grassy water." The early Spanish explorers who landed on the coast of Florida were intimidated by the vast swampland and didn't venture into it. Sight unseen they named it El Laguno del Espíritu Santo. They suspected, as most Spanish explorers seemed to do, that there were vast treasures hidden somewhere in its depths, but the prospect of poisonous jungles and fetid lagoons peopled by ferocious animals and reptiles kept them from exploring it.

It was the English, says Marjory Stoneman Douglas in her history of the Everglades,* who first named them "River Glades" on their maps, which later became Ever Glades. The word "glade" comes from the Anglo-Saxon *glaed,* which means bright or shining. The same word appears in the Scandinavian languages meaning a bright streak or patch of light in the sky. In modern English "glade" has come to mean a little green clearing or meadow in the forest. Whatever the early explorers did intend, today's description of "river of grass" seems best to describe what the Everglades really are. For this is how the Everglades began, as a river of grass and water, a great slow-moving river, as much as 70 miles wide, moving down to the sea.

The water came from Lake Okeechobee, 100 miles to the north, spilling over its southern rim to spread out and out and flow imperceptibly southward. No rush of water here, it was more like a steady seepage, for the land sloped from Okeechobee to the Cape at only one inch in a mile.

* *The Everglades: River of Grass,* written for the "Rivers of America" series and reprinted by Hurricane House, Miami, Fla.

This vast Everglades basin, some 2,700 square miles, is a shallow saucer, its sides tilted barely enough to contain the water, its porous foundation a great fresh-water sponge—the Biscayne aquifer. The constant seaward movement forms a barrier that the scanty elevation cannot provide to the salt-water tides, and keeps the sea from pouring into that vital water table.

Okeechobee—the Indian name means "big water"—is the largest American lake entirely within a state. Once the overflow from its 700-square-mile surface kept the entire Everglades basin saturated. This constantly renewed supply of fresh water nourished an ecology of plant and animal life without parallel. In the brackish zone, where the fresh water meets and mingles with the salt, it was fantastically prolific, breeding the tiny life forms at the bottom of the Everglades food chain and incubating the fish and shrimp that appeared in such profusion.

Lake Okeechobee overflows no more. The Army Corps of Engineers have girdled it with a massive levee 40 feet high and 200 feet wide at the base. How this came about is a tale no more preposterous than many others, once you have seen what can happen, and usually does.

President Truman signed the act creating Everglades National Park in 1947. Less than a year later the park was in trouble.

It was a year of unusually heavy rainfall. Florida's rainy season came early and stayed late. The saw grass savannahs became lakes. Between Okeechobee and the new park, farms built on the black muckland of the drained marsh became soggy swamps again. The small drainage canals dug here and there to carry off excess water were inundated. In October a tremendous rainstorm finished the job. Farms and cattle went under, and the flood spread north and south into Miami, Fort Lauderdale and even parts of Palm Beach, covering an area as large as New Jersey.

Congress heard the cries of the Florida farmers and called on the Army Corps of Engineers, one of whose responsibilities and duties is flood control. The engineers made a survey and came up with a program for flood control, drainage and reclamation of land. Please note: Called upon for flood control, they gratuitously added two more factors, on the well-known principle that wetlands are wastelands and should be drained anyway.

So in 1948 Congress authorized a project that the engineers assured the National Park Service would not damage or interfere with Everglades National Park. In fact, they said, it would help the park, because in dry periods it would be possible to release water into the park from proposed water-conservation areas.

"This plan," said Peter Farb in *Audubon* Magazine (September 1965): "provided for the biggest earth-moving job since the building of the Panama Canal. The Corps might have prevented future flood damage much less expensively—simply by buying up all of the flood-damaged property, at a total cost of no more than $12,000,000 to $20,000,000, and letting it be flooded. Instead, it proposed a project with a cost at least 20 times the value of the entire flooded area. This project originally estimated to cost $200,000,000 has now (1965) soared to $381,000,000—and some knowledgeable observers believe that eventually it will cost more than half a billion dollars."

Essentially the plan consisted of putting a levee around Lake Okeechobee, to contain the water (and eventually to raise its level), and constructing a series of dikes and canals to carry the overflow east into the Atlantic and west into the Gulf of Mexico.

Below Okeechobee and north of the park two huge conservation areas were set aside as water-impounding zones. Conservation Area 3 bordered the park, with dikes separating its water from the park areas just to the south. A major east-west road, the Tamiami Trail, ran over these dikes, and under it were gates that could be raised to let water flow into the park.

This drainage system from the lake turned out to be quite efficient. When Hurricane Donna poured billions of gallons of water on Florida in 1960 the engineers opened the locks of the new canals and swiftly lowered the level of the lake. The water went into the Atlantic and into the Gulf and there was only minor flooding. Which was fine.

But there was another finger in the pie. The engineers were given the responsibility of building the huge project but not the authority to administer it. For this the Florida legislature created a state agency, the Central and Southern Florida Flood Control District, or FCD. This was a five-man board appointed by the governor, with authority reaching into 18 counties—an area larger than Connecticut, New Jersey and Delaware. Theirs was the real power. The only authority the Army Engineers had was to act in a flood emergency, when it was their responsibility to prevent floods by lowering the water level on the lake through their canals. The FCD ruled on how and where the water should be used.

Sharp criticism came from columnist John Pennekamp of the Miami *Herald* (for whom an underwater park is named—the John Pennekamp Coral Reef at Key Largo in the Florida Keys):

> Right after the Flood Control District was set up, the Commissioners began to enlarge their mission and to become a land reclamation bureau. This put the park in competition with the land owners around Okeechobee. Between the farming projects round the lake and the north border of the Park are 60-odd miles, all Everglades country. It is the water that spilled from the lake and then flowed southward into the park that has been retarded and needlessly wasted in run-offs to the sea authorized by the Flood Control District. Four of the five commissioners come from the general Okeechobee area and presumably this accounts for their disposition toward land reclamation and farm protection.

Another *Herald* writer, Nixon Smiley, remarking that, like most agencies created by the legislature, the FCD was responsive to pressure, added:

> But most of this pressure comes not from the general public—the main body of taxpayers—but from groups wanting special works done. For instance, a group of Marin County farmers and grove developers wants thousands of acres drained to protect its trees and crops from floods.
>
> These influential men, working through the proper county governmental channels, present their request to the FCD governing board, and the request finds its way through state channels to Congress, which in turn authorizes the Army Corps of Engineers to study the feasibility of the request.
>
> This study, incidentally, is under way.

The land discussed for drainage was estimated at more than 735,000 acres.

In 1961 a cycle of plentiful rain came to an end and a four-year drought began in southern Florida. Ponds and sloughs in the Everglades began to dry out. Fish died, the alligators burrowed vainly in the mud, the wading birds found no food. Meantime the engineers had completed the job of impounding Conservation Area 3. With its dikes closed, it still was lush with water from Okeechobee, while just the other side of the dike the park baked in the Florida sun, waiting for rain that did not come.

By 1965 the park was dying. The ponds and creeks had dried into cracked mud. The famed Anhinga Trail, a boardwalk into the swamp built so people could see alligators and wading birds close up, overlooked a gummy mire that held the dead bodies of fish and turtles. The birds were scattered, except for the turkey buzzards reaping a bonanza of death. The saw grass had gone tinder-dry, and great fires swept the savannahs, where the water no longer sparkled. Clouds of smoke obscured the sun. "What had been a river of grass," wrote Marjory Stoneman Douglas, "was made in one chaotic gesture of greed and ignorance and folly, a river of fire."

In a desperate effort to save what they could, park rangers used dynamite to blast out " 'gator holes," which might collect moisture and save some of the alligators. They roped the bigger ones and wrestled them into submission and lugged them to water.

By the fourth dry season the birds had left their nests and their young because they could find no food for them. The great flocks of wading birds were decimated.

The wildlife of the Everglades is predominantly aquatic. Fish, shrimp, turtles, alligators and otters actually live in the water all or part of the time, but even the birds—the herons, egrets, ibis and anhingas—are dependent upon the water for their food. Everglades deer are hardly aquatic, but they feed largely upon plants like the water lily.

As the water level dropped, garfish, a staple food of alligators, began to disappear. Rangers found dead gars covered with a tiny parasite, a sucking copepod, whose numbers had multiplied in the drought. The wood ibis, making its last stand against extinction in the park, had lost half its number, down from about 10,000 to 5000.

Four years of drought saw land animals—raccoons and squirrels and rabbits—coming in to replace the aquatic species. The saw grass was yielding in places to trees. Along the southern edge of the park the salt-water tides curled farther inland, and the salt mangroves moved inland with them. The encroaching salt water brought marine predators, such as the sea snail, and caused a sharp drop in the population of shellfish like clams and oysters. The pink Tortugas shrimp, which spends a six-month growing and fattening period in the brackish estuaries of the Everglades and is worth $30,000,000 a year to Florida shrimp fishermen, was threatened by the increasing salinity.

At last, yielding to pressure from the Park Department, from the Audubon Society, from the angry editorials in the papers, and indeed from the appalled reaction of millions of tourists who came to the park and disaster, the FCD yielded. *One* gate in the dike at Conservation Area 3 was opened *one* inch for *one* week. At about the same time the Army Engineers released billions of gallons of water through their canals to the sea. The trickle of water that flowed into the park was absorbed in the tinder-dry saw grass or evaporated before it got very far.

It wasn't feasible, said an official of the FCD, to get water from Lake Okeechobee to the park. And to drain any considerable amount out of Conservation Area 3 would damage the biology of that section, hurting the fishing prospects. Who was more entitled to the water anyway, the FCD wanted to know, humans or birds? The question ignored totally the fact that more water was being dumped into the sea than the park ever required.

The incongruity of permitting this tiny trickle to flow toward the park while 500 times as much went into the sea caused others to question the complete authority of the FCD. Naples City councilman Joel Kuperberg said, "It's madness. We're destroying the very things that people come here to enjoy" (*Newsweek,* October 24, 1966).

But in Collier County, northwest of the park, the Gulf American Land Company was even then draining 200 square miles of cypress marsh—"Building new worlds for a better tomorrow"—and offering lots for sale in its "Golden Gate Estates." A network of drainage canals was siphoning off more water to the Gulf of Mexico, and a "tree crusher," a machine two stories high, weighing 55 tons, was smashing a 20-foot-wide path of destruction through the pine and cypress at every pass.

"The wilderness has been pushed back," said the ad copy. "With calipers and slide rules . . . draglines and dynamite rigs . . . we are literally changing the face of Florida" (*Newsweek,* October 24, 1966).

Adjoining Gulf American's "estates," the Audubon Society's Corkscrew Swamp Sanctuary, containing one of the few remaining rookeries of wood storks and an irreplaceable stand of bald cypress, was losing water. Gulf American's 60-foot drainage canal, undercutting State Highway 846, was pointed toward the Corkscrew Sanctuary like a thirsty hypodermic needle. "It was like pulling a plug in a bathtub," said Philip Owens, the sanctuary director.

But the blundering was apparently not all one-sided. Nixon Smiley, writing in the Miami *Herald*, said if it was necessary to attribute blame to anybody for the park's troubles, both sides should share it. The National Park Service, he said, had done hardly any research to solve the problems that were beginning to harass many park areas and had made little contact with the Army Engineers in spite of the $381,000,000 project being built on two sides of the Everglades National Park. The Park Service had done little work to determine how much water the park needed in critical times or where the water was needed most. And he concluded:

When the Corps of Engineers wanted to extend canals for a distance into the park, to carry and distribute water more efficiently, the plan was turned down.

The drought caused the park to back away a little from its original ideas and recently the Corps was given permission to extend an existing canal, L-67, from Conservation Area 3 along the eastern edge of the park for 10 miles.

This new extension would bring water to Shark River Slough, where the shrimp grew, and Taylor Slough, a showplace for wildlife.

Then, early in 1966, Governor Haydon Burns announced a new $3,500,000 plan for improving a series of canals to bring water from below Lake Okeechobee to the Tamiami Trail levee west of Miami. Since the land was so flat, pumping would be necessary to move the water, and the Miami *News* estimated that a 500-cubic-foot-per-second capacity would be available to the Tamiami Trail gates, as contrasted with the then available 19 cubic feet per second through the same gates.

The Army Corps announced a $400,000 study of the entire flood-control project directed by Congress. Roger Allin, then superintendent of the park, was not altogether happy about some of this, particularly the plan for canals bringing water from adjacent agricultural areas into the Shark River area. The water would be welcome, but it would be contaminated with pesticides and fertilizers. The fertilizers would increase the growth of algae, choke off some of the smaller water courses, and as it died, use up the oxygen needed by the fish. "And when the fish start to die, the birds go next."

But the statement issued jointly by the Army Engineers and the Park Service was an achievement in itself. It made public recognition of the fact that the park was an aquatic area and must be "nourished with sufficient water to provide the environment necessary for the production and maintenance of aquatic plant and animal populations."

In March 1966 Governor Burns announced that when the water in Lake Okeechobee rose above a certain level the engineers would be permitted to pump more water into Conservation Area 3, from whence it presumably could be released to the park. The point that was continuously being overlooked was that the park could not be allowed to dry out before this water was released. The land slope was so small that the water would not move fast enough or far enough to reach all areas (once they had dried out) before it was lost through absorption and evaporation. What was needed was a continuous flow to maintain the hydroperiod, the time of saturation.

Then the governor compounded the ignorance by asking why the Park Service hadn't put a dike across the southern extremity of the park to keep all this valuable water from flowing off into the bay.

Ecology, apparently, was too esoteric a subject for most, and the idea that the fresh water had to keep flowing offshore to keep the salt water from coming in is something still not widely understood.

A year later, in 1967, the situation was approximately this:

Conservation Area 3, with 450,000 acres, retained about 155,000 acre-feet of water. During the first six months of that dry season about 1,000,000 acre-feet were lost to evaporation and plant transpiration. No water had gone to the park from here, nor, for that matter, to the farms east of the park or to the cities on the east coast. About 46,000 acre-feet had reached the park by canal directly from Okeechobee. And the FCD was dumping a yearly average of 656,000,000,000 gallons into the sea in a nightmare misuse of power.

Meantime the Army Engineers had produced a new toy for the FCD—Drainage and Barge Canal C-111, running southeast from Homestead on the eastern edge of the park to Barnes Sound. The purpose of this canal was to serve a planned industrial complex in the Homestead area. Though virtually finished, the canal was not yet open to the bay. An earthen dam—a plug about 50 feet wide and 13 feet deep—temporarily kept out the salt water of the bay.

A sharp fight erupted over Canal C-111. The Park Service opposed opening the dam on the grounds that it would accelerate the bleeding of fresh water from the park, permit salt water to flood into the canal, and not only damage the park but contaminate the essential Biscayne aquifer with salt water, lowering the fresh-water table and sharply aggravating the water crisis for everyone, including Miami and the other east-coast cities.

A geologic survey test of conditions showed that salt water would have flowed inland on 19 out of 31 days in the selected month of November 1966.

The Army Corps of Engineers was not convinced. Oscar Rawls, chief of project planning at Jacksonville, said, "For an area that is only one foot above sea level, there can hardly be a predominantly fresh water ecology. Sea water already has access."

The FCD was equally unmoved. Its own ecologists advised that fresh water, flowing southward, would keep out the salt water. "Let's pull the plug and see if there is any damage" was the attitude. "Hell," said Park Superintendent Allin, "the whole point is that we don't want to have any damage to demonstrate!" (*Newsweek*, October 24, 1966).

John Pennekamp, in his column, had already been conducting his own brand of warfare against the canal. "Don't pull the plug!" was the war cry. Now the Audubon Society stepped in and filed suit against Secretary of the Army Stanley R. Resor and Lieutenant General William F. Cassidy, Chief of the Army Corps of Engineers. The suit charged them with exceeding their authority and with ignoring specific laws to protect the park. The Audubon Society was joined by 28 or 29 other plaintiffs, including fishermen and farmers.

The Audubon Society has a long history of fighting for Florida wildlife. It was the major force in stopping the slaughter of egrets in the time when an egret plume on a lady's hat was a necessity of life. At least one Audubon warden had been killed by poachers. And the society maintained and guarded a number of islands in Florida Bay as nesting sanctuaries for egrets, pelicans and spoonbills.

The suit jolted the FCD, if only slightly. It did not admit that the canal posed any danger to the water supply, but it backed down. On March 19, 1967—one day before it was to be removed—it agreed to postpone pulling the plug.

By March 1969, when the new Secretary of the Interior, Walter J. Hickel, flew to Florida to confer with the new governor, Claude R. Kirk, Jr., the Everglades water crisis had still not been solved. Plentiful rainfall had eased the threat of severe drought such as killed wildlife and vegetation. But rain would never be enough to supply the park, and the specter of another severe water shortage was still very much alive. Said *The New York Times:* "Rather, as Army engineers planned new works to assure industrial and agricultural users of plenty of water, conservationists worry about a share to save the park."

Secretary Hickel's meeting with Governor Kirk was on the subject of alligator poachers in the Everglades. He wanted Kirk's support for a bill pending in Congress to stop the skin trade at its source by making it illegal to possess or ship alligator skins in interstate commerce. But much of the conversation concerned water. The Park Service was asking for 315,000 acre-feet a year (the new park superintendent, John C. Raftery, told the Wilderness Society in November 1968 that this would not be enough) and agreed to share any shortages in drought years with present users. It refused to share shortages with industrial and agricultural users of the future, since their operations were growing steadily larger.

A footnote to this conference: A year later, in March 1970, this understanding was put to the test at a committee meeting in Washington in an exchange between George Hartzog, Jr., director of the Park Service, and Robert Jordan III, general counsel for the Army Engineers. As reported in Jack Anderson's column (New York *Post*, March 5, 1970), Hartzog reminded Jordan of the agreement that the park would not share water rights with future users because Congress had amended the Flood Control Act in 1968 to put the park needs first.

Jordan denied the agreement. He said the agreement was based on a report that "does not contemplate the creation of priorities for the use of water."

The statement left Hartzog aghast. "I'm utterly appalled at what he advised this committee when he said there was no agreement between us," he said. "I am just simply overwhelmed."

And another footnote: Hunters in south Florida voluntarily cut short the 1969–70 deer-hunting season to save the remainder of the dwindling herd. The season was closed by the state at the request of the South Florida Coordinating Council of Sportsmen's Clubs.

Hunting has been permitted since 1952 by the FCD in their two big conservation areas under the usual seasonal rules of the Florida game department. The heavy rains of 1968 flooded the conservation areas and forced the deer to take refuge in the hammocks. Depleting the food supply there, they began to die of starvation and of a disease carried by a parasite called the "barber pole worm." Their number dropped from about 9,000 to 2,500 *before* the start of the hunting season. Wild turkeys were in the same situation.

Sportsmen and the Florida game department asked the FCD to lower the water level in the conservation areas by releasing more water to Everglades Park. The FCD acknowledged the plight of the deer but said it could do nothing, because its duty lay with the efficient utilization of water, not with game protection.

Then, in the early fall of 1968, a small cloud that had formed like one of the summer storms over the Everglades, blew up into a major eruption. On September 18, 1968, the Dade County Port Authority announced ground-breaking on the world's "first all-new jetport for the supersonic age" in the Everglades between Big Cypress Swamp and the park.

The planned airport was to be 39 miles square, big enough to lose in it the airports at Los Angeles, San Francisco, Dulles in Washington and Kennedy in New York and leave space for extra runways. It was to handle the biggest of the supersonic transports, with runways six miles long. The location was just six miles north of the park.

"Having invested vast efforts and millions of dollars on a famous national park," said Anthony Wayne Smith of the National Parks Association, "we now turn all our engineering powers against our own environmental treasures."

The project was supported by the Dade County Port Authority, the Federal Aviation Administration (which made a grant of $500,000 to get it started), four major airlines, Collier County, the state of Florida and a collection of local boosters and land speculators. The Bureau of Public Roads and the Army Engineers were peripherally involved, one for access roads, the other for the needed drainage and distribution of Big Cypress water.

A major access corridor at least 1,000 feet wide was planned to connect the jetport with Miami (and later with Tampa and St.

Petersburg) by means of auto roads and high-speed trains. The Federal Railway Administration announced a grant of $200,000 to study high-speed ground transportation between the jetport and Miami.

"A new city is going to rise up in the middle of Florida," said Alan C. Stewart, director of the Dade County Port Authority. "You are going to have one whether you like it or not."

To the new park superintendent, John C. Raftery, it meant "Slow death. Portions of the park literally face ruination."

Felice Dickson of the Miami *Herald* (August 22, 1969) soberly quotes Stewart as referring to conservationists as "butterfly chasers" and the rare bird species endangered as a bunch of "yellow-bellied sapsuckers." He also insisted that the jetport would create no pollution and no "excessive noise" and that the delicately balanced ecology of the park would not be destroyed.

Point of view is an elusive thing. From Stewart's position his job was to provide transportation for Florida, the fastest-growing state in the Union—probably the third largest by the year 2000. Florida has already gone through many real-estate booms, some of them disasters, but land speculation continues and zoning protection is feeble. Traffic jams are piling up on the roads and airline tickets are almost unobtainable during the winter season.

Yet the pressure is always for more growth and the drums beat—bigger and bigger, this year bigger than last, next year bigger than this. And those most interested in sheer growth are usually those least interested in its social, human or environmental effects. But the physical environment of Florida, says Paul Brooks of the Sierra Club, "is extraordinarily fragile and vulnerable to misuse. In short, the greatest alteration of the landscape anywhere in the United States is being imposed on the area perhaps least prepared to withstand it."

Land in the conservation area was being condemned at $150 an acre, but all around the projected airport, prices were shooting up (land around existing airports in the Miami area start at $35,000 an acre), with speculators already organizing projects to drain the marsh.

The water loss would affect the western arm of the park in Monroe County, and drainage would hurt. But pollution would hurt even more.

Pollution would come from two sources. One, the effluent from the new complex was estimated in the authoritative Leopold Report * to the Department of the Interior at 4,000,000 gallons per day of sewage and 1,500,000 gallons per day of industrial waste. The inevitable growth of commercial establishments—hotels, motels, shops, living quarters for personnel, and so on—would compound this to a major degree.

A heavy use of pesticides and fertilizers in and around the jetport would become necessary. Fertilizers would lead to eutrophication (aging of the environment due to algae stimulation and loss of oxygen), and pesticides would add to the problem of biological magnification (concentration of pesticides in higher levels all along the food chain).

Pollution along the transportation corridor and in the airport would come from auto and truck traffic of a very high order, together with a leakage of oil and chemicals into the surrounding terrain. The jet planes themselves would produce pollution of a high order. A jet plane sprays a gallon of unburned fuel into the air on takeoff; with the airport in full operation, 10 gallons a minute would descend like rain on the conservation area and the park as the planes turned south in their climbout. The Sierra Club estimated exhaust pollutants from the planes, based on 900,000 flights a year, as something on this order:

Carbon monoxide	9,000 to 72,000 tons
Nitrogen oxides	4,150 to 6,000 tons
Hydrocarbons	13,000 to 40,250 tons
Aldehydes	1,000 tons
Particulates	1,260 to 3,250 tons

As for noise, the Port Authority was quite happy about that. There was going to be noise, lots of it, but who was going to complain about it except the alligators? "The Everglades National Park south of the site at Tamiami Trail assures that no private complaining development will be adjacent on that side," said their report. To them the park was a "sound screen." "The Park and Conservation Area 3 are sound barriers in that no human habitation in these areas is anticipated."

* Environmental Research Report prepared for the Joint Committee of the Departments of the Interior and Transportation. Dr. Luna B. Leopold, head of the Environmental Research Team for Interior and the State of Florida, is an ecologist of the Geological Survey.

As usual, the Indians were overlooked in this assessment of "nobody." The area just happened to be inhabited by the Miccosukee Indians, whose ancestral hunting grounds these were. Some of the airline officials were sufficiently aware of them to refer to the training runway under construction as "The Green Corn Dance Airport" after a ceremony of the Miccosukees.

Noise levels of the supersonic jets are on the order of 120 decibels, with the future SSTs estimated by Boeing to be at 122 decibels. The threshold of actual pain is considered to be about 120 decibels. On its maiden flight the Anglo-French Concorde could be heard in villages 20 miles away. "It is expected," said *Aerospace Technology* * "that the Concorde will exhibit sideline noise levels of about 118 PNdB [decibels of perceived noise] according to U.S. engineers, and may show a rather startling 124 PNdB figure during approach." Today's subsonic jets create a noise level at takeoff of 120 decibels *three miles away.*

But aviation spokesmen could say, "Favorable noise environment of the 39-square-mile site is indicated by large undeveloped areas, with Indian reservations, Everglades National Park and state water conservation areas serving as buffers."

The Leopold Report envisioned another problem, and even its official language carried a chilling impact:

> A severe bird strike problem (airplanes hitting birds) may develop within the airport boundaries, over Conservation Area 3, and in the quadrant southwest from the training strip. This problem would involve large water birds, including several rare and endangered species at altitudes ranging from ground to 2,000 feet. Small animals that seek refuge on the runways in flood periods will add to this problem when they are crushed and attract carrion-eating birds.

The effect of a large bird being drawn into a jet engine on takeoff and suddenly cutting the power of a heavily loaded plane climbing upward need hardly be described.

By September 1969, just before the appearance of the Leopold Report, *Aviation Week & Space Technology*, a trade magazine, was reporting growing opposition to the jetport. Transportation Secretary Volpe and Interior Secretary Hickel had met with Florida Gov-

* Reported in *Sierra Club Bulletin*, July 1969.

ernor Kirk and given their opinion that the site in the Cypress swamp should be disapproved.

Nevertheless, the magazine reported, "the Dade County Port Authority feels confident that the decision it reached during a two-year period while the site was being selected will be favorably reflected in studies which are being conducted to evaluate the site and its potential commercial use." Said potential was given as 50,000,000 passengers and 1,000,000 aircraft operations annually—higher than the Sierra Club estimate.

Evidently the Port Authority did not expect at least one of the studies referred to—the Leopold Report—to characterize the jetport as a disaster. Richard H. Judy, deputy director of the Port Authority, said they had worked with or consulted with three Federal and five state agencies, as well as regional and local officials, the Air Transport Association, major airlines, Governor Kirk and Chief Buffalo Tiger of the Miccosukee Indians, and had received written approval of the site from all. Twelve alternate sites were considered, including four in the Everglades, before the one on Miccosukee hunting grounds was selected.

But in an interview with James Malone of the Miami *Herald* (September 5, 1969), Nathaniel Reed, conservation adviser to Governor Kirk, said, "We have found that there are many alternative sites, in Collier, Broward, Hendry and Palm Beach Counties, which the Dade County Port Authority never fully investigated. Some of them include state-owned land that could be acquired without cost."

He went on to say: "It seems the jetport as it is now planned would just do an extraordinary amount of damage. They [Dade officials] were warned well ahead of time that this might happen. I can't think of picking a worse spot for an airport."

Richard Judy said he could not believe a decision on Federal backing of the site—without which it could not be built—was likely before the completion of a locally sponsored study aimed at determining if the jetport could be made compatible with natural surroundings. "I think this is just another part of the conservationists' propaganda campaign," he said.

By now the training runway, two miles long, had been completed, with bulldozers smashing through Bloodhound Hammock and draglines gouging out mountains of dolomite from the cypress swamp for fill. To make this runway optional had required an initial expenditure of $13,000,000, of which the Federal government had contributed $750,000.

But the opposition was now in high gear. The Leopold Report had had a strong effect on Governor Kirk, Secretary Hickel and Secretary Volpe. A coalition of 23 conservation and economic organizations had been formed in Washington, with the Audubon Society, the National Parks Association, the Wilderness Society and the Sierra Club in the forefront, and these had made a strong protest to Secretary Volpe.

At a hearing of the Senate Committee on Interior and Insular Affairs Senator Gaylord Nelson of Wisconsin had said, "Moving the jetport will cause one hell of an uproar, but it can be done." And in a slap at the land developers he said, "You can grow as long as you can steal water from the park."

In an apparent reference to the fact that projects in south Florida were being financed with 80 percent of Federal money and only 20 percent of state money, he said, "We don't have to sit here and be clobbered by the state of Florida. We can either stop the jetport or admit publicly that we are going to destroy the park."

The Leopold Report said it bluntly: the jetport "will destroy the south Florida ecosystem and thus the Everglades National Park." And the report added:

> The south Florida problem is merely one example of an issue that sooner or later must be faced by the nation as a whole. How are the diffused but general costs to society to be balanced against the local, more direct, and usually monetary benefits to a small portion of the society? Concurrently, the society must ask itself whether the primary measure of progress will indefinitely be the degree of expansion of development, such as housing, trade and urbanization, even at the expense of a varied and, at least in part, natural landscape.

The decision came on January 13, 1970. At a joint press conference at the White House, Secretary Hickel and Secretary Volpe released a statement from President Nixon (who was not there) announcing an agreement with state and local authorities forbidding the construction of a jetport at the chosen location. President Nixon's statement said in part: "The agreement affirms the need to conserve our national heritage; it does not deny the need for new airport facilities in Florida. The Government will cooperate in finding ways to create such new facilities without new threats to the environment."

The agreement, however, permitted the use of the single runway (which the Leopold Report said in itself would be disastrous) until a new site should be found. "When such a regional airport site has been acquired, priority will be given to the construction of training facilities and, when these facilities have been completed, all training operations conducted at the present airport will be transferred to that site."

What does that mean? It means the runway continues in use, with its measure of damage and pollution, until the Dade County Port Authority and the Federal government agree upon a site—at no cost to the Port Authority. That means the U.S. taxpayer makes a gift to Florida of the $13,000,000 the Port Authority sank into the original site.

Moreover, as Joe Browder of the Audubon Society has dourly observed, no issue holds the public attention forever, and if the search for another site can be prolonged sufficiently, the climate in Washington could well change and permit the Port Authority to go quietly ahead with its plans.

"So long as the training airport is in use," said the Leopold Report, "pressures and plans for its expansion will continue and will inexorably and surely lead to ecosystem destruction completely."

Just such a suspicion led two men who met in the Everglades, one from Vermont and the other from Ohio, to file suit in the U.S. District Court in Miami to stop operations at the airstrip. Said one, "They just tossed us a bone. Finding a new site may take several years. In that time, much of the flora and fauna will die off and the place will be a garbage dump" (*New York Times,* February 3, 1970).

Predictions like these used to sound like scare psychology, or as Richard Judy of the Port Authority put it, "conservationists' propaganda." But they no longer do. It turns out, said one newspaper columnist recently, that those alarmist nuts were right all along. A new boom psychology seemed to have been set in motion by the jetport furore, and speculators who bought large tracts of Big Cypress still have visions of draining it and selling it by the foot. More than 50 square miles of the Everglades above the park in private ownership are scheduled for drainage, including Gum Slough, a major water-supply basin. The Audubon Society has gone to court again to try to stop this drainage project.

North of Lake Okeechobee the Kissimmee River, one of the last wild rivers in Florida, is being obliterated. A monster machine, 100 feet high and 200 feet long, is literally devouring the river, spewing out mud, water, plants and trees, obliterating the curves and leaving behind it a straight ditch like a monument to a robot society.

Peter Matthiessen writes in the *Audubon* Magazine of March 1970:

> The ditch is straight, as rigid and ugly as a caved-in subway tunnel and approximately as useful. On one side is a wasteland of gray-green ooze where spill from the ditch, dumped over high raw sterile spill banks, has buried the bass and turtles and alligators of the marsh; the bird legions are gone, leaving only the black requiem species—the crows, vultures and grackles—to pick over the dead and dying plants and animals. On the other side, the marshbed, sucked dry, has turned a funereal black, as the clean water of its ecosystem, increasingly precious as populations grow and resources decline, pours away to Okeechobee and the sea.

Ask about this project, says Matthiessen, and the Army Engineers refer you to the FCD. The FCD refers you to the Army Engineers. And there is talk of "feasibility" and "water management." Meantime the marsh is drained and a few landowners will own some thousands of acres of dry land, gift of the U.S. taxpayer. Other landowners are saying that the entire Big Cypress area—half a million acres—must be drained in the name of progress. At, of course, public expense.

If the drainage system around Lake Okeechobee has already cost $200,000,000 and is only half completed, it would appear that the Audubon Society is right in saying that it would be cheaper to buy all of Big Cypress than to turn the Corps of Engineers loose in it. "It would cost the American taxpayers more to let speculators develop the Big Cypress," says Joe Browder, Audubon field representative, "than it would to buy the whole area and preserve it." If this is not done, he says, "hundreds of millions of tax dollars will be spent on similar land reclamation projects in southwest Florida—and the Big Cypress and Everglades National Park will be lost" (*Audubon* Magazine, March 1970).

As the Biscayne aquifer serves to supply water to the East Coast, Big Cypress provides water for the cities on the West Coast. Drain it, and southwest Florida will be destroying its own water source—to the peril of the expanding populations of Fort Myers and Naples.

The Biscayne aquifer is in danger, too, from salt-water intrusion as Miami's increasing demands lower the fresh-water table. The Biscayne aquifer is huge, as much as 200 feet thick in places, and one of the most prolific water producers in the world. But, as John Pennekamp said, years ago,

> This great aquifer is remarkable not only for size but for its fragility. The danger of salt-water intrusion comes from big machinery which might scratch through its skin, that developers and harbor deepening enthusiasts might use to break into the aquifer. We had a lesson in that when the Miami River was deepened to fifteen feet. For a mile inland from the riverbanks the fine old wells turned salt. It's man, not the age-long wetness of the park, that is the menace to the aquifer.*

An engineer with the Dade County Health Department, Peter J. Baljet, forecasts a water shortage in Miami by 1985. The present consumption of 230,000,000 gallons a day can rise to 1,400,000,000 gallons if population growth reaches the 4,000,000 figure predicted.

Like the park, Dade County (which is mostly Miami) depends on the FCD for water, which is to say Lake Okeechobee. The peculiarity of the lake is that although large, it is shallow. Engineers figure that its usable water lies in a three- to five-foot-deep area. To deepen it by raising the level of the lake poses a risk that is tied to Florida's hurricane-prone location. Okeechobee is a dangerous place. A hurricane can drive millions of gallons of water over the shallow banks and cause frightening floods. The 1926 and 1928 hurricanes in the Okeechobee area drowned 2,500 people. This is why the Army Engineers were given the project of ringing the lake with a levee so its level could safely be raised by two feet. This project was budgeted at $50,000,000, but Congress has yet to appropriate the money for the final two years of work.

Each foot the dike is raised will add about 150,000,000,000 gallons of capacity to the lake. Two feet will meet the needs of 4,000,000 people, but not, in addition, industrial requirements, the needs of the farmers and the Everglades National Park. So, although billions of gallons are now wasted to the sea whenever the lake level gets too high, eventually demand will overtake capacity.

The Army Engineers estimate that in the next 30 years consumer

* Quoted in *American Forests,* July 1966.

demand will increase four times and agricultural demand 50 percent, but agricultural volume will be four times that of the population demand. They have therefore recommended adding an additional four feet to the dike around Lake Okeechobee, with additional canals and pumping facilities, at a further cost of $74,000,000. This, they believe, will hold the line until the year 2004.

It would appear that the aim of government in this country is to satisfy the largest possible number of competing demands. In practice this means government yields in the area of greatest pressure, too often without a clear idea of the greatest good. In theory, the Army Engineers, the state of Florida and the Department of the Interior have all accepted Everglades National Park as an equal partner in the disposition of south Florida's water. But in practice, what happens when the pressures mount, as they will again? In March of 1970 a three-day conference on the Everglades in Miami brought a gloomy prediction from Dr. Robert C. Harris, Florida State University oceanographer. "The Everglades are going to be gone," he said, "whether it's the year 2000 or 2100." Pesticides, pollution and unbridled drainage have imposed a "death sentence" on the marsh that is turning it into a desolate wasteland.

The process has already begun for Big Cypress. The oil-prospecting crews have moved in with their equipment and the search for oil is on. As the waters recede, the drills bite through the hardening swamp muck, and the great white birds fly in panic at the roar of engines spelling finale to a million years of quiet.

The insatiable urge to manage and develop the environment is not only at odds with but is outraged by the simple suggestion to leave it alone. Leaving it alone for esthetic reasons would be reason enough if man were as highly developed as he likes to think. But esthetic reasons offend the practical people, who suggest that conservationists be given a bucket of butterflies and a net to keep them quiet.

The role of wetlands in this country has yet to be widely understood. As a stabilizer of the water supply, as the most prolific incubator of life, as a refuge for wildlife, as a resource rather than a wasteland—these roles, as a spokesman for the Dade County Port Authority once said, are all right, "but it's not the way the great American system operates."

"When all our wells have turned to salt," wrote Al Volker in the

Miami *News,* "when every stream stinks, when Biscayne Bay is an open cesspool, when all the Everglades has been paved, when the air is black with soot and possibly carcinogenic hydrocarbons, then we will be poor indeed, even if such despoliation has made us all millionaires."

Florida is lucky. That state of affairs has already arrived for a good part of the nation.

If we are unable to save and preserve the Everglades, we lose something more precious than even water or birds. We lose the promise of mankind itself.

PART II

The Woodlands

CHAPTER

4

The Forests

A COMMON misconception about evolution is to think of it as teleological—that there is purpose in nature's design and that the direction is ever upward and onward. It would be nice if it were so, but it isn't.

Evolution is genetic adaptation to the forces of environment, and it can be downward as easily as upward. There are plenty of gloomy observers who believe man is deteriorating mentally and physically, not to say morally.

We tell ourselves we have come a long way from the brutalities of the Dark Ages, yet the twentieth century is the bloodiest in history. We tell ourselves we have better ethical standards than the robber barons of the nineteenth century, who manipulated the stock market and ravaged the land, and that these things could not happen today. But couldn't they? Or have we merely learned to cover such activities with a more sophisticated patina of words—to use the big lie more expertly?

Perhaps there are no villains, only men with the capacity for believing that what they want is right. And perhaps this kind of tunnel vision is the real problem.

In the early part of 1969 a bill supported by the lumber industry, the National Forest Timber Supply Act, was introduced in Congress.

It was presented simultaneously in the House as HR 12025 and in the Senate as S 1832.

Three excerpts from this bill provide its essence:

Sec. 2. The Congress hereby finds that in order to meet increasing national demands for lumber and other wood products, including that needed for home construction, it is necessary to increase substantially the timber yield from the commercial forest land of the nation, including that in the National Forests . . .

Sec. 3. As used in this Act the term "commercial forest land" means forest land which is producing or is capable of producing crops of industrial wood and not withdrawn for timber utilization by statute or administrative regulation.

Sec. 7. The Secretary of Agriculture . . . shall immediately establish programs to carry out the policy and purposes of this Act and shall specifically:

(1) develop into optimal timber productivity as soon as possible the National Forest commercial timberlands

(2) revise the allowable annual harvesting rates in National Forests to take into account (a) rotation ages estimated to be appropriate for markets and technology at the expected time of harvest (b) the need for and benefits from use of high level current harvest rate options available within sustained yield limitations, and (c) increased timber yields which will result from application of the measures authorized by Sec. 6 of this Act, as rapidly as possible after such measures have been undertaken . . .

The legalese doesn't sound too ominous. And, in fact, when the bill first came to light in the House Agriculture Committee it aroused little interest in Congress and the press. But some voices began to be heard.

The *Morning Tribune* of Lewiston, Idaho, in the heart of the western timber country, which would be most affected by the bill, called it (May 22, 1969) "potentially mischievous legislation," the sort of measure that a lumbering area could be expected to support. But, said the paper, it would force the Bureau of Land Management and the Forest Service to increase the allowable cut, in some cases without regard for such other considerations as watershed protection, erosion prevention and the preservation of fish and wildlife habitat. "It thus represents the very opposite of multiple use, a concept to which the lumber industry has expressed consistent devotion."

There were no guarantees, said the *Tribune*, that the additional lumber would go into low-cost housing, and added:

The bill's chief defect, however, is its failure to recognize that the public lands, including the forests, belong to all the people and not only to the loggers. In its present form it would reverse some 60 years of forest conservation policy by denying the legitimacy of any use incompatible with the harvesting of timber. The Lewiston region, which depends upon its forests for many things besides logs, should be particularly wary of the National Timber Supply Act of 1969 as it is written now.

Although not the only issue involved, multiple use of public lands became the hinge of the debate in Congress. Meantime the opposition was getting organized, bringing together such diverse groups as the National Wildlife Federation, the Sierra Club and the National Rifle Association. It may come as a surprise to some that the last-named has always considered itself a conservation organization.

Support for the bill created even stranger bedfellows, namely, the lumber industry and the housing industry. Home builders, traditionally hostile to timber producers because of the high prices for wood, had been convinced by lumber spokesmen that prices would come down if increased supplies from the national forests were made available.

When conservationists began calling the proposal "a crass raid on a natural resource" the proponents of the bill worked a quick change of title. It now became the National Forest Timber Conservation and Management Act of 1969. The word "conservation" in the title obviously removed any unpleasant connotation it might otherwise have had.

But on May 13, 1969, *The New York Times* had commented acidly:

The timber shortage scare is unconvincing and the proposed solutions are equally suspect. There is no certainty that the additional timber sales ordered by Secretary Hardin will make more lumber available for low-cost housing. Moreover, the dislocation in home building is not primarily caused by lumber prices but by tight credit. The cost of money, not of lumber, is the real problem. With lumber company profits at record-breaking levels, the industry is suffering no hardship.

The Nixon Administration, therefore, appeared to be in a position first of keeping credit tight and interest rates high, then of pushing a bill in the name of a solution that appeared to be no solution at all.

Both Secretary of Agriculture Clifford M. Hardin and Secretary of Housing and Urban Development George Romney approved the bill, and Mr. Hardin wrote to Congressman W. R. Poage of Texas, chairman of the House Agriculture Committee, urging its enactment.

The essence of the bill was to increase cutting in the national forests. Before examining the consequences of such a move, it might be instructive to ask why the lumber companies were so interested.

The contention was that a shortage of timber began in 1967 and caused prices to rise precipitously. Between March 1967 and March 1969 softwood lumber climbed 66 percent and softwood plywood 124 percent in price. *U.S. News & World Report* on April 21, 1969, quoted a lumber-industry prediction that supplies would be about 8,000,000,000 board feet short of the nearly 50,000,000,000 board feet needed to meet housing goals set by Congress.

On March 19 President Nixon ordered an increased cutting of 1,100,000,000 board feet from national forests over a period of fifteen months. Said *U.S. News & World Report* on April 21, 1969: "In the long term, the large increases of timber that are needed to meet the fast-growing demand for lumber in the U. S. are going to have to come from the national forests. On that point there is broad agreement among forestry experts in Government and in private industry."

Private industry said the wood must come from the national forests because that's where it was—60 percent of all the nation's remaining softwood. They disliked this talk of "raiding" the public forests, because timber was a resource to be used, not just to stand there, benefiting nobody. Nature, they believed, left to herself tends to be wasteful and profligate. With "more intensive management, accelerated growth and increased timber production," the forests could be made to yield greater "harvests" indefinitely without any harm.

The obvious question that arises went unanswered. If the timber industry is so good at better management and accelerated growth, why weren't they using it on their own vastly greater lands? The commercial forests in the public domain comprise 97,000,000 acres—only 19 percent of the 510,000,000 acres of commercial forest land in the country. Why did the lumber companies have such a disastrous shortage on their five-times-greater acreage so that now they needed this small remainder to bail them out?

One reason given—a short-term reason—was that heavy snows,

combined with a scarcity of railroad cars and a dock strike, had tied up shipments. But a more unpleasant, long-term reason was that the industry had been overcutting for years, in spite of all the talk about good management and sustained yields.

An equally unpalatable fact was that 90 percent of the national forest is open for cutting anyway. The new bill merely called for an increase in the *rate* of cutting. Logging in the national forests has increased from about 5,000,000,000 board feet a year in 1950 to nearly 12,000,000,000 board feet today. The Department of Agriculture is selling timber at a rate of about 50 percent more than can be sustained by replanting.

Testifying before the House Committee on Agriculture in May 1969, Edward C. Crafts, former assistant chief of the Forest Service said:

> For many years in connection with its periodic reassessments of the timber situation, the Service has predicted a prospective shortage of softwood sawtimber and this is exactly what is facing the industry now and why it is turning to the National Forests as its own lands have been depleted of mature timber.
>
> It is my feeling that the Forest Service has been, and is being, pushed dangerously close to the brink with respect to timber management on the National Forests. I do not believe in brinkmanship when it comes to depleting the natural resources of the United States.

An even more curious fact emerged as the debate developed. The great lumber shortage was not a real shortage at all. The lumber industry, while pleading scarcity, was exporting logs to Japan—4,000,000,000 board feet in 1969, almost 20 times the amount exported in 1960. According to *American Forests* (August 1968), Weyerhaeuser, one of the larger timber companies, sold Japanese buyers 160,000,000 board feet of logs in 1966 and 220,000,000 in 1967. Georgia-Pacific, Crown Zellerbach, Scott, Simpson, and St. Regis are other companies reported selling large volumes of logs to Japan. "It appears somewhat ludicrous," remarked William E. Towell, executive vice-president of the American Forestry Association, "to be demanding increased harvest of National Forest timber to meet lumber shortages without plugging this export drain" (*American Forests*, June 1969). Mr. Towell added:

> By law these forests are to serve "multiple uses" for the benefit of all

citizens. Certainly timber supply is one of these uses, but not the only one. We can afford to increase sawlog production only if adequate safeguards are taken to protect water supply, recreation, wildlife, wilderness and all the other benefits that Federal forests provide. Even an acute shortage of home building materials cannot be used as an excuse for upsetting the "multiple use" principle of public land management. Let's get our Federal forests up to full production, but not at the expense of long-range benefits to all citizens whose needs are far greater than lumber and plywood.

There is not only no real shortage of lumber, there is no increasing demand, as the industry maintains. In 1960 the actual demand for wood products was about the same as it was in 1910, and the demand for lumber was actually *down,* in spite of the fact that the population had doubled and the gross national product was five times higher. The per capita consumption of wood was 219 board feet in 1950 and 170 board feet in 1968.

Wood substitutes, plastics and aluminum, are being increasingly used in home building, and it is the fear of this increasing use that is a major factor behind the industry's drive for more and cheaper wood. Moreover, as land grows scarcer and more expensive, the trend of builders is toward multiple-family units, which use only two-fifths as much material per family as individual homes.

Lumber prices were already sagging from their 1968 and 1969 highs as the bill went to Congress. Douglas fir two-by-fours, which were $150 a thousand board feet in March 1968, were $89 in January 1970, and plywood sheathing, which was $160 a thousand square feet in February 1969, was $82 in January 1970.

Obviously the industry, concerned over the trend toward wood substitutes, was looking for greater volume and cheaper sources to tilt the balance back toward the use of wood. For surely the very companies that cry "shortage" loudest are the last to suggest we use less wood, are the most aggressive in promoting greater uses of paper, disposable tissues, throwaway paper clothing and similar high-waste products. There is no thought of conserving, but always an aggressive campaign to use more and sell more.

The costs of intensive management mentioned in the Timber Supply Act—fertilizing, spraying, pruning, and development of hybrid species in the national forests—would be borne not by the timber industry but by the public. The price the companies would pay the government for the trees would not reflect these costs, so that the public would be subsidizing an industry whose profits are already at an all-time high.

Federal Trade Commission figures indicate that the lumber industry showed the highest profit of any industry in the United States in 1968. Profits rose 97 percent before taxes and 91 percent after taxes, for a total net profit of $635,000,000. In 1968 seven companies made new records for net income, with earnings per share up 50 to 100 percent. Typical examples: Georgia-Pacific's net profit was $76,000,000; Weyerhaeuser made $105,000,000. For 1969 Georgia-Pacific's net profits were up another 19.7 percent, with the company clearing $91,760,000 on sales of $1,160,160,000.

All of which may have prompted the Louisville *Courier-Journal* to write on November 4, 1969: "What is being attempted here is a steal, at a hideous potential cost to future Americans."

Lumber-company spokesmen are traditionally contemptuous of wilderness advocates who wish to "lock up" valuable woodland to keep it from being "developed." To legislate wilderness, say the lumbermen, is to perpetrate a fraud on the American people. It is stupid not to "harvest" available timber or to administer public lands for an increase in harvest capability. The industry does not disguise its opposition to the adding of more land to parks and wilderness areas.

Ralph Hodges, general manager of the National Forest Products Association, is quoted as hinting (*New York Times*, February 5, 1970) that the next step might be to seek permission to cut in areas now reserved. He has said, "Salvage and improvement cuttings within parks and other reserved areas could be made without impairment of scenic and recreational values."

A timber raid on the parks was exactly the next step predicted by conservationists as both sides lined up early in 1970 for heavy lobbying in Washington.

The Agriculture Committee had approved HR 12025 with only one dissenting vote and two abstentions. One of the abstainers was Congressman W. R. Poage of Texas, who was having misgivings and who said on February 6 to a reporter, "Frankly, I'm going to vote against it, as I see the bill now. I think it opens the door to destroy multiple use."

Ten conservation groups used the office of Congressman John D.

Dingell of Michigan as their headquarters and began mailing out literature in opposition to the bill. The lumber people had made even more elaborate preparations for their campaign.

A major voice for the industry, the American Forest Products Institute, was deodorized with a new name, the American Forest Institute. This is the public-relations arm of the industry, with four news bureaus, an education department and a research department. It publishes literature, distributes photographs, and sponsors stories and editorials for newspapers.

Other associated members of the industry publish their own papers, the largest of which is the *National Timber Industry*.

Another group, called Outdoors Unlimited, professes to be recreation-motivated, but its connections are with the timber industry, and it has been active in campaigns involving parks and wilderness areas.

A new group proposed during the hearings on the bill by the National Forest Products Association was to be called FACE, for Facts, Action, Communication and Evaluation. This group was intended to work at local levels to defeat legislation proposed for the protection of wilderness or scenic areas.

In opposing additional parks or wild areas, the timber industry has traditionally used the principle of multiple use as an argument. Now came a change. A brochure issued by the National Forest Products Association said that "precisely the point" of HR 12025 was to tip the balance of multiple use toward increased timber production.

On February 21, 1970, Senator Edmund S. Muskie of Maine, chairman of the Senate Subcommittee on Air and Water Pollution, joined the Sierra Club in accusing the Nixon Administration of ignoring its own Environmental Policy Act in its zeal to see the timber bill enacted.

The Environmental Policy Act, signed by President Nixon on January 1, 1970, requires all Federal agencies, when recommending legislation that affects the quality of the environment, to submit a detailed statement on "the environmental impact of the proposed legislation." This statement is to be sent to the President and the Council on Environmental Quality and is to be made public.

Senator Muskie and Lloyd Tupling of the Sierra Club said that neither Secretary Hardin nor Secretary Romney had complied with this requirement, and Senator Muskie asked whether the studies called for had in fact been made. Although both Messrs. Hardin and Romney had endorsed the bill, neither Cabinet official had submitted a statement on the environmental effect of the legislation. Following Senator Muskie's statement, neither Cabinet officer was available for comment, but Mr. Hardin's executive assistant, E. F. Behrens, said that such a statement would not be "germane" (*New York Times,* February 22, 1970). The bill, he said, reaffirms the multiple-use concept and doesn't change the functioning of the Forest Service except to make its funding more complete. "The bill has no environmental effect."

As more searching questions were asked, support for it gradually eroded. *The New York Times* asked editorially on February 25, 1970, why 30 of the 33 members of the Agriculture Committee had voted for the bill based on the need for housing, but only nine of the 33 had voted for the Housing and Urban Act in the first place.

Party lines were crossed as the attack on the bill was led by Congressman Dingle of Michigan, a Democrat, and Congressman John P. Saylor of Pennsylvania, a Republican. Congressman Saylor said the bill contained "one rabbit for the people and one horse for the timber industry." Conservation-minded congressmen had misgivings about cutting, in the next 15 years, an amount of timber that originally had been planned to be spaced out over the next 100 years.

Michael McCloskey, conservation director of the Sierra Club, wrote in the *New Republic* of December 13, 1969:

> At a time when ecologists are warning of the potential for disaster in the unstable nature of highly intensive agriculture, it is the height of folly to apply the same techniques to the national forests. These forests, found often on highly mountainous terrain, include many sites that are already ecologically fragile. The forests serve a multitude of purposes in addition to forestry: watershed to produce pure water, habitat for large stocks of native wildlife, pastures for grazing, and they are laced with tracts valuable for recreation. The sensitive practice of multiple use precludes their being treated as farm lots under monoculture. Monoculture will invite increasing insect attacks, prompting repeated resort to insecticides damaging to wildlife and fish. Fertilizers will leach into pure mountain [water] supplies. Forest scenery will disappear as a large percentage of the canopy is removed to make way for young row crops of conifers. The richness of a forest that is more or less natural will be supplanted by a relatively sterile and unstable ecology.

Two things became increasingly apparent as the House stalled through February on taking action with HR 12025. One was that the principle of multiple use was evidently in danger despite protests to the contrary. The other was that the bill contained a very insidious menace in that it actually empowered Congress to change the classification of 97,000,000 acres of public forest, making it in effect a farm for high-yield forestry. No Congress had ever attempted to decide the predominant land use on so large an area of the public domain.

On February 26 the House, by a roll call of 228 to 150, refused to consider HR 12025. It did not vote down the bill. It merely refused a resolution to bring it to the floor for debate.

The victory for conservation, such as it was, also was short-lived. In June President Nixon sent an order to the Secretary of Agriculture and the Secretary of the Interior "to permit increased harvest of softwood timber consistent with sustained yield, environmental quality, and multiple use objectives." The executive order, said Congressman Saylor, achieved by fiat what could not be done legislatively.

Conservationists reacted angrily. C. R. Gutermuth, vice-president of the Wildlife Management Institute, said (*New York Times,* June 26, 1970), "They [the timber industry] have been cutting the hell out of their own lands [to maintain exports to Japan] and depending on the Government to bail them out by permitting excessive cutting in the national forests."

The President's order said that increased cutting should be "consistent with sustained yield—there must be money for manpower to plant new trees." But as for allocating the money—a wide back door was left open as the statement continued: "Any additional funding required for the execution of these plans will be reviewed by the Bureau of the Budget in relation to over-all national priorities."

Upon which Spencer Smith, Jr., of the Citizens' Committee on Natural Resources, remarked, "I am a little restrained about approving the program [for cutting] before seeing the money in the budget [for reforestation]." And conservationists pointed out that the Forest Service was already 5,000,000 acres behind in its normal reforestation program.

The Sierra Club said (*Sierra Club Bulletin,* July 1970): "Nixon's major justification for boosting the cutting rate is the current housing shortage and the assumption that the decline in new housing starts is due to a lumber shortage. Yet the nation's lumber mills are now overproducing and more than 30 billion board feet of standing timber, almost three times the annual cut, has been sold to lumbermen, but remains on national forest lands."

As of now the executive order stands.

In one form or another, accelerated cutting, on the gamble that forest improvements will come later, will continue to disturb ecologists. Gordon Robinson, consultant to the Sierra Club, says that the quality of forestry in the United States is low and getting lower, since industry continues to cut timber faster than it grows.

Good forestry practices would ensure that cutting be done on a long rotational basis, from one hundred to two hundred years. This would maintain a mature forest, in contrast to the industry's pressure for a revision of annual harvesting rates to allow for rotation ages of different kinds of trees. In practice this means cutting trees at a younger age than good silvics (the ecology of trees) would approve.

Cutting, says Mr. Robinson, should be done frequently and lightly, with no more than 10 percent of the available trees removed at any one time. If clear-cutting—removing all trees from a section—must be done, the opening in the forest should be kept as small as possible, presumably to minimize wind throw and water erosion.

Maintaining a mature forest greatly reduces fire hazard. The full canopy of leaves provides a controlled climate, cool because of the shade, moist because of the transpiration of the leaves. Mature trees are relatively free of branches close to the ground and their thick bark is heat-resistant. In such a climate, fire is slow to start and finds little to feed on.

A textbook illustration of these principles occurred in 1956 in Klamath County, California, when a fire went through three sections of woodland. One section was national forest, recently logged in accordance with the Forest Service rules, which allowed removal of about 60 percent of the trees, mostly the largest specimens. The second section was virgin timber, not logged. The third was private land, on which a lumber company had cut everything permitted by law, amounting to 90 percent of the trees, leaving only those under 20 inches in diameter.

The fire roared through the private land and destroyed every remaining tree. It killed about half the trees left in the national forest section. In the virgin forest it burned only a few patches of new growth and a few dead trees, leaving virtually no permanent harm.

Timber companies that practice clear-cutting argue that this is the most economical way to harvest the logs and they replant anyway. In too many cases they replant with trees of one kind only—all growing in machinelike straight rows like corn. This is particularly notable in the West and South, where thousands of acres of former cotton fields are now planted in fast-growing pines for pulpwood. A pure stand of trees, like a pure stand of corn or wheat, encourages the invasion of specific insects or infection by fungi or disease, whereas a mixed crop doesn't offer such fertile breeding grounds for an epidemic. Consider a specific fungus of one kind of tree. In a pure stand of these trees the air-borne spores will find a resting place wherever they land, while in a mixed stand most will never find a host.

As for clear-cutting, some recent studies have turned up unexpected contradictions. A U.S. Forest Service study in the Redwood Purchase Unit showed that it cost $11.45 per thousand board feet to clear-cut an area, and $11.37 per thousand board feet to log it selectively.

In a pine forest, clear-cutting took 133 man-minutes per thousand feet, heavy selective cutting 118 man-minutes, and a light sanitation cut 119 man-minutes.

Taking only the largest trees also afforded the greatest handling efficiency. It appears to cost twice as much to fell, limb, load and cut into boards a tree 12 inches in diameter as it does a bigger tree, 30 inches or more.

Clear-cutting has other penalties as well. If heavy rains follow, they cut the ground into gullies, wash away the topsoil, uproot remaining trees and may start floods or mudslides. Lacking rain, and with the natural canopy removed, the moist forest floor dries out, humus is destroyed, the population of small insects, mites, other arthropods, worms, fungi and bacteria die out, and the entire ecology of this vital few inches is damaged, often beyond repair.

It is for this reason, this fatal alteration of the forest floor, that second-growth timber is never the same as the original growth. The new trees are much more likely to die of disease before they attain the growth of their forebears, and the soil, reduced in fertility, never seems able to produce the magnificent specimens it once did.

A tree is one thing, a forest another. It is more than likely that the timber operator who sends in a chain-saw gang to destroy a virgin forest enjoys as much as anyone the stately beauty, the cool shade and fragrance of trees around his home.

If man's progenitors descended from the trees to walk upright (and the construction of our hands with an opposed thumb argues a development for grasping branches), it seems to have left us with mixed racial memories of the forests once our home.

Early man personified the trees and peopled the forests with spirits good and bad—nymphs, dryads, gnomes, trolls, pixies, leprechauns, witches, satyrs, fauns, banshees, and several minor gods. Wood nymphs lived in trees and perished if their abode was cut down.

In many people an elemental fear of the deep woods lingers, perhaps related to a fear of the dark. The shadows conceal evil spirits, or at the very least, wolves and bears.

Mythology is filled with tales of wood spirits that cast spells on man and tormented him with every manner of cruel trick. Zeus spoke to man through the leaves of a great oak in the sacred grove of Dodona, and the oak tree was sacred also to Thor, god of thunder. The groves were sanctuaries where hunted men were safe, or where oracles foretold the future.

Norse mythology was born in the forests. The universe of the Norsemen was supported by a great ash tree named Yggdrasil, whose roots went through the three worlds and even to Asgard, the dwelling place of the gods.

> Three roots there are to Yggdrasil
> Hel lives beneath the first.
> Beneath the second the frost-giants
> And men beneath the third.

At the foot of Yggdrasil sat the three Norns: Urda the past, Verdandi the present, and Skuld the future. The Norns "Allot their lives to the sons of men, / And assign to them their fate." *

There were sacred groves in Africa and Borneo, in Timor and

* *Mythology* by Edith Hamilton, Little, Brown & Co., and Mentor Books.

India, in Sumatra and the Philippines, and in China, where the ginkgo tree was saved from extinction only because it was sacred.

The first clear-cutting was done by missionaries taking literally the Bible's admonition to stamp out idolatrous worship—"But ye shall destroy their altars, break their images and cut down their groves."

They didn't stamp it out completely. While the sacred groves at Uppsala in Sweden may have lost their capacity to awe, Valborgsmassoafton, the arrival of spring, is still celebrated with bonfires and dancing on April 30, and Midsummer Day in Norway is a holiday, with roots deep in the love of forest and earth.

But the great apostle of the forest was the American Indian, the Indian of the woodlands. Deeply religious, he personified everything—animals, birds, trees, water, mountains and sky—and wove all into a moving story of creation. The Indian had a song or prayer or ceremony for every act and duty. He was so much a part of nature's design that he asked forgiveness of the animal he killed for food. And when his forests were gone, he had nothing left but his song of the ghost dance:

> Pity me, Wahkonda.
> My soul is ever hungry.
> There is nothing here to satisfy me,
> I walk in darkness;
> Pity me, Wahkonda!

If to primitive man the great forests were often fearsome places where gods and spirits lay in wait for him, modern man remains almost as ignorant of the meaning of a forest, of its complex and vital ecology.

A forest is an ecosystem of innumerable living things dominated by the trees.

A single acre of woodland may contain only 150 or so mature trees, but there will be about 10,000 saplings and shrubs, ten times that many herbs or grassy plants, at least 500,000 earthworms, certainly a million or more soil mites, trillions of fungi—and when we get into the area of nematodes, actinomycetes and bacteria, we run out of numbers. It has been said that to step on a forest floor is to cover a million living organisms.

The trees continually modify and temper the environment of this system. The full canopy of leaves intercepts and diminishes the wind, filters the sunlight and controls the humidity. The constant falling of leaves, twigs and branches provides a continual shower of organic matter—about two tons an acre annually—which provides the home and the nourishment of the soil organisms below.

If this rain of material from the trees was not quickly broken down by fungi, beetles, springtails, termites, earthworms and bacteria, the forest would soon fill up with its own debris. But the litter is rapidly reduced to humus, which forms a springy, organic mat perhaps six inches or so thick, in which live most of the small organisms that carry on the essential work of returning plant tissues to the basic chemicals of the soil. This mat is continuously cultivated, plowed, turned, aerated and digested by its inhabitants.

An indispensable workman is the earthworm, which aerates the ground with its endless tunneling, grinds up bits of plant matter, passing them through its digestive tract with soil, and leaves castings everywhere that are vital in building earth. Darwin estimated that earthworms chew up enough litter and soil to produce as much as 18 tons of castings an acre, and this may be conservative. A naturalist in Africa watching through a rainy season came up with a count of 107 tons an acre.

Nor are earthworms alone. Ants, crickets and termites make their own tunnels, and bigger creatures—shrews and moles and mice—dig for themselves.

The common millipede, like the earthworm, lives on decayed vegetable matter and, with many kinds of insects, chews leaves and litter, breaking them down so bacteria can then digest the sugars and starches in the plant tissue, releasing carbon dioxide. The carbon dioxide is used by the trees and other green plants for photosynthesis, which works essentially like this: Water is brought up from the soil by the plant roots to the chlorophyll in the leaf. Utilizing the energy of sunlight, the chlorophyll splits the water molecule into its components, oxygen and hydrogen. The oxygen is given off by the leaf as a by-product; the hydrogen is combined with the carbon dioxide taken from the air and converted to sugar.

This complex process is probably the most important single activity on earth, because green plants are the primary producers upon which all other forms of life depend. Any creature is either a consumer or a decomposer, but none can produce its own food from air and water. The base of the food chain on land, just as in the sea, rests on green plants.

Of the consumers some—insects, rabbits, deer, woodchucks, even squirrels, which live on nuts and seeds—can digest plant tissues and convert them into protein. The carnivores digest vegetation poorly or not at all, but can assimilate protein directly from meat. These are the predators—the foxes, owls, hawks, cats, wolves and snakes, which prey on the plant eaters. A few, like man, raccoons and bears, are omnivorous and will eat almost anything.

There are checks and balances in this system so long as it is in balance itself. Insects are normally held to a reasonable number by birds, lizards, frogs and other insects. The rabbits are controlled by hawks, foxes and owls; the deer by wolves or puma. Eliminate any class of predator and you invite a population explosion among the animals that are its natural prey. Upset the balance by planting to a single crop and the particular insect or mite or fungus that grows best on that crop is vastly encouraged, sometimes to the point of no control. Spray insecticides that cannot tell friend from foe and whole species of insects good and bad are wiped out, with residues left to imperil all the rest of the food chain.

First settlers in America found virgin forests that were quite different from the second or third growth of today. An old forest is a climax forest, meaning that it has reached a state of equilibrium dominated by the species of trees best adapted to that particular biome or habitat. Habitats, smaller sections of a biome, can be quite localized, varying on a single mountain according to differences in elevation, exposure, sunlight, wind or soil, so that even a climax forest is made up of patches of different trees rather than a solid stretch of one kind. Much of the northeast forest was a mixture of deciduous trees and conifers—hemlocks, white pines and hardwoods.

Even when lumbered, burned, or destroyed by hurricane, the regrowing forest will, over a very long period of time, tend to reestablish itself in the original climax pattern. As the tide of agriculture moves west, abandoning the stony fields of New England, the trees begin to come back in a predictable pattern.

First come the weeds and shrubs, then the pines. White pine seeds are light and are broadcast on the wind by the millions. Some sumacs, wild cherry, and birch compete for a while, and the birches play a definite role in sheltering the pine seedlings, but at last they are gone and the pines dominate.

They dominate too well, growing so thickly that they shade out their own seedlings. Then the oaks and maples, which will grow in the shade, take hold, and by increasing the shade, gradually crowd out the young pines. As the old pines die off, a second forest is formed of hardwoods.

The dense shade of the oaks and maples now makes it difficult even for their own seedlings to grow, and the newest sprouts are hemlock, beech and basswood. The oaks and maples die of age, injury or insect attack. They are replaced by hemlock and beech with a scattering of assorted hardwoods, and the equilibrium is reached in which only the dominant species have much chance. The climax forest is restored.

Each stage of this process, which takes hundreds of years, affects not only the trees but the rest of the web of life. Ornithologist Roger Tory Peterson did a study on bird succession in Maine. It showed that while a recently cutover area still grew only weeds and shrubs, it harbored birds, such as the Savannah sparrow and the bobolink, adapted to sun-baked, weedy pastures. As brush grew taller, field sparrows, towhees and a few warblers, such as the Nashville and chestnut-sided, moved in. With the first trees came new birds—redstarts, vireos and ruffed grouse. As the pines thickened, more warblers and the first thrushes appeared. And when the dominant spruce forest of Maine became established, the birds included woodpeckers, kinglets and many types of wood warbler.

A similar progression of animal species can be expected, each inhabiting the niche where the food on which it depends can be found.

Climax forests can be disturbed and altered by a variety of disasters—fire, flood, earthquake, or disease caused by viruses, fungi or other parasites. Diseases have been encouraged by some of man's manipulations: intensive lumbering, plantings of a single tree species, importing of foreign trees that may open a Pandora's box of infections for which there are no native controls—diseases like the oak wilt, the Dutch elm disease and the chestnut blight.

Imported insects also create havoc. A well-publicized example is the gypsy moth, brought over in 1869 by a French scientist who expected to revolutionize the silk industry. Some of them got away from his Massachusetts laboratory and started their own population explosion. Defying efforts at control, the moths have spread through New England into New York, New Jersey and Pennsylvania. Gypsy moth caterpillars will cover the trees in season like a thick blanket of hairy spaghetti, stripping the leaves clean. White pines may die after such complete defoliation, although most trees recover very well after one such onslaught.

Forest fire under primitive conditions is usually not a catastrophe. Climate control beneath the full canopy of a mature forest tends to limit the fire, and the big trees are often no more than singed. But fire in a logged forest is something else. It has not been the practice of loggers to clean up after their operations, and a slash of branches, bark, chips, sawdust, broken stumps, trampled brush and crushed saplings remain like a scene of battle. The slash is perfect fuel for a roaring inferno that will generate enough heat to consume anything, especially since the canopy has been opened to permit updrafts.

Such a fire can create a blackened and desolate burn that leaves a graveyard of charred stumps for decades. There are thousands and thousands of acres of such burned areas in the country. The Peshtigo fire in Wisconsin and Michigan in 1871 burned 3,780,000 acres and killed 1,638 people. According to ecologist Peter Farb, more than 200,000,000 acres have been burned over, some of it so badly that by 1920 about 80,000,000 acres were useless even for crops.

Fires in the United States are fought more efficiently now, but a current danger area is thinly populated Alaska. In the summer of 1969 alone more than 4,000,000 acres of forest burned in Alaska, posing the question of effect on its ecology, on future floods, soil erosion, watersheds and wildlife.

A serious fire, even if not causing total destruction, burns off the humus, kills young growth and opens the way for different kinds of trees to come in. Thus Minnesota and Wisconsin today have forests of jack pine, a tree that had a very small role in the original climax forest. Logging removed the great white pines and hardwoods, fires swept the cuttings and prepared the way for a proliferation of jack pines, whose tight cones need heat to open and scatter their seeds. Similarly, current logging of Douglas firs in the Pacific Northwest is not encouraging regrowth, for hemlock and cedars are coming in instead.

An experiment in the Kaibab forest of Arizona backfired when Theodore Roosevelt made it a national game preserve in 1906. To build up the deer herds, government hunters opened a campaign against predators, killing puma, coyotes and wolves. From an estimated 4,000 deer the herd increased to 100,000 in 1924 and ate themselves out of house and home. They ate every leaf and twig they could reach and then died of starvation by the thousands.

Sixty percent of the herds were estimated to have died in two winters. By 1942 their number was down to about 8,000. Meantime the plant ecology of the Kaibab had been shattered. Willows and wild raspberries were all but gone; aspen had been damaged for years to come; firs, spruce and pine had been drastically set back, and the balance had been tilted to an invasion of grasses and herbs, which, if it had continued, would have completely altered the character of that area.

By contrast, an example of balance is the Isle Royale National Park in Lake Superior. A herd of moose, living undisturbed on the island and increasing in number, had begun to overbrowse the foliage and create the same situation that had been fatal to the Kaibab deer. But then some wolves arrived, probably coming over the ice in winter. Now a stable community has been achieved. The moose population is stationary at a healthy number, with the wolves keeping the herd limited to the vigorous individuals by culling out the aged and sick. The herd is improved, and both moose and wolves maintain a balance of numbers, since the wolves cannot increase beyond the availability of their own food supply.

For a long time people were not able to grasp the essential fact that ecology means balance. Nor could they see that forest ecology was the controlling factor in watersheds, in holding the delicate line between supply and flood. The aerated texture of forest soil will easily hold 14 times as much water as soil in an open field. The experience of two towns in Utah—Farmington and Centerville—both on the Great Salt Lake at the foot of the Wasatch Mountains, is an illustration.

Drenching rains had all but buried Farmington in mud and rocks that came sliding down from the mountains. Highways, farms, orchards and homes were buried in feet of mud. Each summer the sudden heavy storms of the area repeated the disasters. No one seemed to know why. The rain did not appear to be heavier than in other years. Even more baffling, it was not happening in nearby Centerville, which was receiving the same amount of rain.

Investigation finally made it clear that the mountains behind Farmington had been stripped of tree cover by overgrazing and by fire. At Centerville the cattle had been kept out of the mountain forests, and the fires had been controlled. Runoff measurements showed one stream in Farmington carrying 160 times the water of a similar stream whose watershed had not been cropped bare.

Once the problem was understood, a replanting program was undertaken in Farmington; cattle were barred from mountain slopes and fire control instituted. Since 1936 there have been no floods in Farmington.

Admittedly, all this sounds most chaotic. The human hand appears so heavy that destruction seems to follow wherever it rests. Can a forest be managed at all—in the parlance of the industry, can it be "harvested" on a continuous basis without its ecology being destroyed?

A fascinating clue is in the story of Juriens, a village in the foothills of the Jura Mountains in Switzerland.

No one in Juriens pays a tax. Village taxes are paid to the canton and the Federal government, but all the money comes out of the village treasury. And all the revenue for the village treasury comes from the Juriens forest.

It is not a big forest—about a mile and a half square. With 400 inhabitants, it comes out to about three and a half acres per person. Yet the logs from this little patch of woodland have kept Juriens tax-free and debt-free for hundreds of years.

Swiss forestry law says, very simply, that the forests must not be reduced. Whoever cuts a tree must see that another replaces it. A government forester marks each tree that may be cut, and each year the village officials decide how many need be cut. If the village does not need as much income one year, fewer trees are cut. There is never an attempt to cut more trees than needed merely to *accumulate* income. And since the Federal and canton taxes are calculated on the basis of income, the tax will actually be lower the years of lighter cutting.

Selective cutting permits the forest to reseed itself; there is no need for artificial planting. All rubble is cleared away, no underbrush or slash is left to create a fire hazard, and the woods are more like a park than a jungle of wild growth. The people of Juriens picnic in their woods, ski or walk in them, or sometimes just sit there.

Pierre Fraley, writing in *American Forests,* quotes a verse found on a forester's cabin wall in Juriens:

> In the deepest woods
> Lies the heart of our country.
> A people without forests
> Is a dying race. That is why
> When a tree perishes we grow
> Another on its grave.

There are other villages in Switzerland like Juriens where for hundreds of years the forest has paid the taxes and furnished employment and income to the residents. Apparently it can be done.

A different situation exists in the British Isles, where only patchy woodlots remain of the once-great forests of oak and beech. In Scotland overcutting and overgrazing have reduced the forest area until today only 9 percent of the land is wooded. The Forestry Commission in Scotland has projected a replanting program of about 36,000 acres a year, so that by 1975 there should be an increase of about 340,000 acres. Private planting is said to be going on at a rate of about 15,000 acres a year. In Scotland, Wales and England the Nature Conservancy manages 95 national nature reserves, most of them small.

Germany's famous Black Forest has undergone changes in the direction of management and speeded growth. Once a climax forest of native hardwoods, the character of the Black Forest has been altered by the introduction of fast-growing softwoods such as the American Douglas fir. Steep hillsides in the Black Forest are now cultivated in fast-growing conifers for pulpwood.

In Norway, where forest products are the second-largest industry, there was no real forest policy until the period of World War I, when heavy cutting caused some alarm and a survey of resources was undertaken. Until then there had been no thought given to conservation or replanting. It was only because of Norway's small population that the forests were protected from serious overcutting. Today Norway's modern foresters are talking, in a familiar tone, of bringing their forests to full production. As part of the program, they are bringing in foreign species from America, notably Sitka spruce and hemlock.

Norway is mostly mountain, with a very long and exposed coastline on the Atlantic and Arctic oceans. The mountain forests protect against erosion, and the forests on the coast provide a shield against the wind. A rise in temperature along the coasts for the past few years has encouraged replanting with a view to establishing new forests to replace the "overage" trees growing there now. It is hoped that 1,800,000,000 trees will have been planted by 1975. Continued planting of spruce is making it the dominant forest tree in many parts of Norway.

Scandinavian forestry is strongly influenced by cooperatives. The owner of a small woodlot can bargain with a big lumber company from a position of strength because he is part of a movement that includes about three-quarters of the farmers and forest owners in these countries. In Sweden forests owned by the big lumber companies are mostly in the north. But the South Swedish Forest Owners' Association, with 43,000 members, mostly farmers, owns 5,000,000 acres of woodland, the biggest sawmills in Sweden, two paper mills, two pulp mills, two particle-board plants, a packaging plant and a factory for making prefabricated houses.

While wood, wood products and paper are generally thought of as the produce of the forest, with some fringe products, such as turpentine and maple syrup, the chemists of our bigger companies are busily wringing an increasing variety of new materials from wood.

One estimate is that at least 2,600 silvichemicals have so far been identified and are being extracted from wood chips, bark, sawdust, or even the fluid left after pulping operations. Material once discarded or burned is now worth more than $260,000,000 a year to wood processors.

West Virginia Pulp & Paper Company estimates revenues of about $12,500,000 in 1964 just from chemicals extracted from the pulp liquor left from papermaking.

Research in silvichemistry is funded at better than $20,000,000 a year and gaining. Industrial chemicals already being produced in quantity from wood include acetic acid, methanol, propionic acid and methyl acetone.

Vanillin, a synthetic flavoring much less expensive than real vanilla, was once widely sold, but for some time it (along with a number of food dyes) has been regarded with suspicion by nutritionists as a possible carcinogen.

A kind of molasses processed from wood sugars is being promoted for cattle feed, and another chemical dinner for cattle made of sawdust is also being developed.

Lignin, one of the essential components of woody tissue, has endless possibilities since its molecule is built on a benzene ring, the basis for hundreds of organic chemical compounds. Crown Zellerbach Corporation is making phenolic compounds from lignin that go into paints and varnishes.

One of the more dramatic materials extracted from lignin is dimethyl sulfoxide, familiarly known as DMSO. This substance created much excitement a few years ago when it was found to relieve miraculously the pain of bursitis, arthritis and migraine headache. Just touched to the skin, DMSO penetrated the entire body so rapidly that its odor could be detected minutes later on the breath. It had the unique ability to carry in with it whatever medication it was paired with, and these might be many since it is a powerful solvent. But even by itself it appeared to relieve the pain of arthritis and migraine like magic.

The interest in DMSO led to considerable uncontrolled experimentation with commercial grades (as opposed to medicinal grades), and the bubble burst when it was found that laboratory animals had developed cataracts. Research was precipitately halted but DMSO has undeniable possibilities that cannot be ignored, and it is inevitable that a more closely controlled research will continue. Meantime industrial-grade DMSO remains in such demand that Crown Zellerbach has undertaken a large addition to its Louisiana plant to raise capacity from 5,000,000 pounds a year to 8,000,000 pounds.

Other materials processed from lignin include phthalic anhydride, a major intermediate for paints and plastics; dimethyl sulfide, an odorant for natural gas (so that leaks will be apparent by smell); a drug called quercetin, and a number of waxes and similar chemicals.

Each in his own sphere, scientists of the various disciplines are groping on the edges of a forest ecology they haven't even tapped. While chemists look for new products to be derived from wood, biologists probe for a better understanding of life and death, health and disease.

The great broad-spectrum antibiotics that have saved millions of lives—the family of tetracyclines, streptomycin, and chloramphenicol—come from soil organisms originally discovered in a teaspoonful of earth. Scientists search in virgin areas, such as the Amazon, where few men have ever been, for new clues to life processes.

The expedition of the scientific vessel *Alpha Helix* to the Rio Negro in Brazil in 1967 produced enough material for scientists to work on for years, enlisting biochemists and physicians from America, Brazil, Norway, France, Britain and Germany. The studies in neurobiology, plant and insect physiology, evolutionary changes in fish, and deep-sea physiology may have yielded clues—obscure, but clues—to cancer, to hormone substitutes for pesticides, and to new hallucinogens useful in studying the mechanism of the nervous system.

Such studies are a logical extension of the search for drugs from plants which has gone on in the forests and jungles for years and has already produced some of our most valuable remedies. They include reserpine, quinine, aspirin, digitalis, and many more. Margaret Kreig, in *Green Medicine,** estimates that about 47 percent of prescribed drugs are from plant sources.

For thousands of years men have cut down trees with no understanding of the havoc that would follow. The sterile and eroded areas around the Mediterranean, once lush and fertile, are evidence enough. In America, with a tradition of fortunes made overnight by exploiting natural resources, there have been few brakes and little condemnation of the exploiter. But there are hopeful signs.

The awakened conscience of a Congress refusing to act on the timber bill is one sign. A rising question as to the need for being bigger and better, producing more and more, consuming more and more, is another. We have been geared to a way of life that encourages more consumption, more waste. But now the question is being asked: How do we stop waste, how do we keep from being buried in our own garbage?

The forest is a good place to start. We can stop being so profligate with wood. We can recycle many wood products—paper, for example—and we can encourage chemists to find uses for the materials thrown away or burned.

* Rand McNally, 1964.

Above all we can recognize our own place in the chain of life and know that we are part of an ecology, not superior to it, not detached from it.

In China, long ago denuded of much of its forest cover, millions of people are today planting hundreds of millions of trees, for the first time in centuries. A militant dictatorship can call out a labor force any time it likes. We need only to know the facts.

And one more thing we need. We need to remember that a forest is a place of enchantment, a place of life. Because it offers beauty, solitude, freedom and silence, it becomes increasingly vital as these become harder to find. The more complex and artificial our civilization becomes, the more the simple verities will be needed.

To walk in the woods, to see a deer, to hear the birds and smell the fragrance of growing things is a renewal, a contact with life that no one should be denied. For this alone, to re-establish our contact with earth, it must be perpetuated. And beyond that, we know now that when it goes, we go too.

CHAPTER

5

Drama in the Redwoods

DRIFTING in from the Pacific, fog blankets the northern California coast, to condense like soft rain on the slopes where the redwoods lift a prodigious canopy to the sky. The redwoods are children of the fog, their crowns cooled by mist, their roots moist in the groundwater so near the surface.

These hills and valleys absorb water lavishly. As much as 55 inches of rain fall a year, and this is trebled by the near-constant drip of mist, condensing as sunset cools the air.

Only here, in the perpetual gloom and wet, do the giant redwoods flourish. The coastal redwood, *Sequoia sempervirens,* is the tallest tree in the world. The loftiest specimen, discovered in Redwood Creek Valley by the National Geographic Society in 1966, is 385 feet high, roughly equivalent to a fifty-story building.

The impressive size of the redwoods has always inspired a tragic dichotomy of reaction in viewers. One group sees the giant trees as unique and irreplaceable and wants them preserved as part of our national heritage. The lumbermen, seeing each tree as an incredible bonanza of boards, looks upon the redwood as a cash crop. The conflict that has erupted between these two groups has gone on for a decade. It grew steadily in bitterness and vituperation to a noisy climax in the fall of 1968.

In October of that year President Lyndon Johnson, as one of his last official acts, signed Public Law 9-545, authorizing a Redwood National Park. A great many well-intentioned people heaved sighs of relief that the redwoods were saved at last.

Alas for illusion. The bill was inadequate to begin with, but even what it authorized is still largely a myth, existing mostly on paper, with reason enough to fear that it may never amount to very much.

From the beginning the struggle over the redwoods involved a great deal of money, a fair number of jobs, and a precious "right" of private enterprise, which is perhaps best summed up in the words of one entrepreneur, who said, "They're my trees, and by God I'll cut 'em down if I want to." Where money is concerned it is very difficult to offer as an opposing argument the rights of generations yet unborn to see a redwood forest. The issues, in fact, were so many, and so complicated, that it seems important for the record to sort out and briefly chronicle the events of the past few years.

The conflict was primarily between a number of conservation groups on one side and the lumber interests on the other. The conservationists simply wanted to save as many of the great trees—some of them 2,000 years old—as possible. They proposed to do this by combining and enlarging some of the 28 scattered state parks in California into a National Redwood Park, although the plans favored by different groups varied considerably.

The lumber companies fought the idea of a national park in principle and resisted giving up any of their lands for the purpose of incorporation into such a park.

Looking at the proposals themselves, at one extreme was the Sierra Club's plan calling for the largest park area, about 91,000 acres. However, it was not the acreage as such that was important here. The Sierra Club had clearly defined reasons for the location and type of redwood forest it proposed to include, and just as clearly defined objections to the areas included in some of the other proposals.

The Sierra Club is a national organization of about 100,000 members, with headquarters in San Francisco and chapters around the country. It is generally considered the most active and aggressive force in the conservation movement. But the Sierra Club did not consider its plan for a redwood park extreme at all. It pointed out that it was essentially the same proposal originally made by the

Department of the Interior's National Park Service (from which the National Park Service later retreated), based on a survey of watershed boundaries needed to protect the areas containing the finest trees. Moreover, said the Sierra Club, what is so extreme about a proposal that would have saved only 2 percent of the virgin trees still in private hands?

The lumber companies, swinging some weight in Congress, opposed the establishment of a National Redwood Park on the grounds that the state of California was perfectly able to protect its own parks and the best trees were already saved anyway. The industry offered 8,000 acres containing prize trees to the state and also proposed opening their lands to the public for recreational purposes (presumably hunting and fishing), making available for this purpose a total of 365,000 acres.

The National Park Service, forced to back away from its original proposal, then offered a series of six alternate plans, gradually scaling down from the original approximately 90,000 acres to a mere 43,000 acres. Even more serious than the smaller size, from the Sierra Club's point of view, was the fact that the final proposal, which came to be considered the Johnson Administration Plan, omitted some of the areas the Sierra Club considered essential to a successful park.

Between the Sierra Club and the administration fell a miscellany of other groups: the National Audubon Society, the National Parks Association, the Save-the-Redwoods League, the Wilderness Society, the American Forestry Association, and a number of others. Some of these backed the Sierra Club while others endorsed the Interior Department's smaller park with reluctance but with the conviction that it was better than risking defeat in an all-out fight against both the administration and the lumber companies.

By the end of 1966 there were 20 different park plans. But by spring 1967 these had been boiled down to three, generally considered to be: (1) the Sierra Club Plan for a 91,000-acre park based on the Redwood Creek area; (2) the Johnson Administration Plan for a 43,000-acre park in the Mill Creek–Del Norte area; and (3) the "Livable Park and Seashore Plan" proposed by Congressman Don H. Clausen of California.

The Sierra Club Plan was embodied in a bill introduced in the House of Representatives by Congressman Jeffrey Cohelan of California as HR 2849 and in the Senate as S 514 by Senator Lee Metcalf of Montana and 16 Senate cosponsors, including Robert Kennedy.

The Johnson Administration Plan, backed obviously by President Johnson and the incumbent Secretary of the Interior, Stewart Udall, was endorsed by the Save-the-Redwoods League, which had favored the Mill Creek area for a park.

The Clausen plan was considered to be favored by the lumber companies (if any) and by Governor Reagan of California. It was criticized as merely linking three existing state parks with a narrow conservation corridor, with little or no acquisition of additional virgin redwoods.

Anthony Wayne Smith, president and general counsel of the National Parks Association, appeared before the Senate Subcommittee on Interior and Insular Affairs (in one of dozens of similar hearings) to speak for a bill, S 2962, which would have created a national park joining three areas—the Jedediah Smith Redwoods State Park, the Del Norte Coast Redwoods State Park and the Mill Creek Basin—and then go on to link up with the Prairie Creek Redwoods State Park and the Redwood Creek Basin to the south.

The intent of S 2962 was to combine the Sierra Club and administration plans. As Mr. Smith explained:

> There has been much controversy as to the relative merits of these two proposals. The National Park Service recommended the Redwood Creek region originally, and then changed its position to favor the Mill Creek region. The fact is that both areas are of national park caliber and should be included in the project. The two areas would be joined by a substantial corridor along the coast between the Del Norte Park and the Prairie Creek Park. This combination would give the American nation a Redwoods National Park worthy of the name; nothing less will do so.

President Kennedy delivered the first special message to Congress on conservation in 1961; since then it has become an annual event. President Johnson's 1967 message, "Protecting Our Natural Heritage," made his project plain. "We must preserve a significant acreage of these primeval redwoods as a national park," Johnson said. "This is a last chance conservation opportunity."

The words "last chance" did not seem an exaggeration, considering that virgin redwoods were being felled at the rate of 10,000 acres a year and the Sierra Club warned that cutting was proceeding in areas planned for inclusion in the new park. Such cutting was endangering the watersheds of areas supposed to be already under protection.

The original expanse of coast redwoods was estimated at 2,000,000 acres. Of this, 1,270,000 acres had been cut over. There were some 50,000 acres in state parks (with another 50,000 acres of surrounding lands) and about 275,000 acres of virgin redwoods remaining.

Congress had felt the pressure from the administration and from vocal conservation groups even earlier. By September of 1966, with a mounting outcry against the continued felling of some of the large and prized trees, Secretary Udall and a coalition of congressmen had persuaded the lumber companies to call a moratorium on cutting in those areas that were being considered for a national park.

Senator Henry M. Jackson of Washington, chairman of the Senate Committee on Interior and Insular Affairs, announced that legislation concerned with a Redwood National Park would be the first order of business for 1967. But on January 14 he reported that hearings would be postponed until April 18 at the request of Governor Reagan, who wanted more time to study the problem.

What had been a small private war between the California Redwood Association (the lumber companies) and a few conservation groups became national and loud. It spread to noncombatants. Such completely disparate publications as *Life* Magazine and *Chemical Week* picked up the cudgels in their February 1967 issues.

Wrote *Life*, arguing that the compromise administration bill was the only one with a chance of winning both congressional and California state appproval: "We urge Congress to pass it—and at this session. No more time should be lost in wrangling, or else, at the rate redwoods are being felled, the whole question may become academic and the earth, as Thoreau mourned a century ago, may indeed have been made 'bald before its time.' "

Chemical Week, a trade magazine read by executives in the chemical industry, and traditionally sympathetic to business, nevertheless chided the lumber companies for their "dubious arguments" and urged "that the companies involved make more than a token gesture to save our redwood heritage, that they help in selecting a site that will best serve the interests of the nation. And we urge Congress to act swiftly, and justly compensate the companies for any losses incurred. To do less would incur an irretrievable loss."

Before we examine the arguments themselves let us stop a moment and consider Exhibit A in the case. What and where is a redwood?

There are two species of trees in California generally called "big trees," "redwoods," or "sequoia." They are related but distinctly different species.

The coastal redwood, *Sequoia sempervirens,* grew originally in a great belt, as much as 100 miles wide, south from the Oregon border to just below the San Francisco peninsula. This tree grows at relatively low altitudes—sea level to about 2,000 feet—along the moist, foggy shoreline.

Sequoia gigantea, or Sierra redwood, grows only in the Sierra Nevada Mountains, in smaller, more isolated groves than the great stretch of coastal redwoods. While not as tall as *sempervirens,* it is a more massive tree. The specimen known as General Sherman is only 272 feet high, but its trunk has a diameter of 34 feet and a circumference of 101 feet, 7 inches. It would take 17 men with outstretched arms, fingertips just touching, to encircle it.

The weight of such a tree is anyone's guess, but lumbermen have estimated that the General Sherman would supply more than 600,000 board feet of lumber, or enough to build 35 houses of five rooms each. Obviously such a tree is a commercial prize, but *gigantea,* because of its relatively small numbers, was early believed to be a vanishing race and won early protection. It does not, therefore, face the same danger as the coastal redwood.

A riddle that these tall trees offer is how they lift water from roots well below the ground to the topmost leaves, a distance that may exceed 400 feet. This is no small engineering achievement, since a quick whirl with the slide rule shows that to raise water say, 450 feet, requires a pressure of 210 pounds per square inch. Add friction between the water and the walls of the tubes in which it travels, and pressure may need to reach 420 pounds per square inch.

A man with a hand pump can put about 15 pounds of pressure on a column of water and raise it about 30 feet. A power pump will obviously offer more than 15 pounds pressure, but the real problem is that if you start to add more pressure you simply pull the column of water apart. How then does a sequoia manage to do more than

10 times better without power and without breaking up the column of water?

The answer, botanists believe, is in the cohesion principle. Water molecules will cling together with great tenacity if they are tightly enclosed in a narrow channel. The narrower the channel the greater the tensile strength of the water column. The tiny tubes through which the sap rises in a tree trunk are so fine that theoretically they could lift a column of water to a height of 6,500 feet—which leaves even a sequoia room to grow.

But what creates the lift? The leaves (or needles, since the sequoia is a conifer) lose water by evaporation and other processes, which creates a difference in pressure above and below. This moves water into the lower channels and then pulls it aloft. There is a pressure differential between the water in the leaf and the water in the roots of 300 to 400 pounds per square inch.

The rise of water in a tree has been measured at nearly 150 feet an hour. As for quantities, redwood statistics are not available; but a palm tree in a desert oasis may pull up 100 gallons a day, just to keep even with evaporative losses.

The redwood has other distinctive features. The soft, reddish wood is virtually immune to decay and insect damage. Redwood trunks lying on the forest floor remain sound and strong for years after an ordinary log would have been reduced by fungi and bacteria to a spongy ruin. A natural resin in the wood makes it distasteful to termites and other insects—another reason for its popularity as a building material.

The redwood tree has so thick a bark that it is relatively undamaged by an ordinary fire. Hence forest fires are actually its ally, clearing out brush and competing species and maintaining optimum conditions for the redwood. Lightning may shatter a redwood's crown—and often does, since the trees are so high—but it will survive. It is, in fact, almost indestructible, the evidence being that there are many specimens 2,000 years old or more. The tree has only one real enemy: man. Even a redwood cannot survive a chain saw, although the stump will often send up new sprouts.

Curiously enough, the great size and age of the trees might lead one to believe that they are slow-growing. Actually they are among the fastest-growing trees known, although they mature slowly and centuries are required for them to reach their full, unbelievable size. A young sequoia grows very rapidly at first (several I started from seed in small pots grew 18 inches high in two years), but in its sixties the pace of growth slows and it may then take 500 years to reach maturity.

Although fossil specimens indicate that redwoods have been around for about 100 million years, the tree as we know it has been growing in California for about 20 million years. The first European to see the redwoods was Don Gaspar de Portola, a Spaniard leading an expedition up the coast of California in 1769. The great redwood belt was undisturbed, however, until the California gold rush began in 1849.

The story is told of a pioneer wagon party that came upon a grove of redwoods and stopped to stare in awe at the unbelievable, towering trunks. Then one man said, "Wouldn't it be a sight to see one *fall?*"

The idea caught on like fever. They broke their journey and set to with hand axes. They worked in shifts, day after day, hardly eating or sleeping, exchanging one axeman for a fresh one as quickly as he tired, gnawing their way like persistent worms into the huge trunk. And finally, after days and days of feverish chopping, the great tree began to groan and tilt, and then it came down, all the way from its majestic height, smashing aside the smaller trees in its path and raising a great cloud of dust as the earth shuddered under the impact, and the men yelled themselves hoarse in excitement.

They had no use for the tree. They just wanted to see it fall.

In the 115 years that followed the California gold rush, lumbering began and increased. Trees that were already towering giants at the time of Christ were felled at an increasing tempo—300 acres a year, then 3,000 acres a year, now 10,000 acres a year. Before 1980 increasing mechanization will make it possible to cut 30,000 acres a year, with only one-sixth or one-seventh of the original virgin acreage remaining.

A Sierra Club advertisement in *The New York Times* of January 25, 1967 explains it thus:

> The arithmetic goes this way: at present 85% of the two million virgin acres has been cut; 3% of the original virgin acreage is held in a number of small California state parks, while the other 12% that's left is scheduled for cutting, which would make a total of 97% of the redwoods given over to that purpose. A Redwood National Park at Redwood Creek would

save, in one forest, an additional 2% of the virgin growth as well as a lovely remote beach area, a number of spectacular wooded hills where redwoods are displayed in the variety of growth conditions in which they thrive, and a navigable river which includes the Emerald Mile, a stretch of huge redwoods running along both sides of the stream. The net effect then, would be that instead of 97% of the original redwoods going to cutting, only 95% would be gone, and we would then have a real sweep of forest large enough for people to walk in without it seeming like a parking lot outside a baseball game.

This advertisement opened with an unashamedly emotional headline:

HISTORY WILL THINK IT MOST STRANGE THAT AMERICA COULD AFFORD THE MOON AND $4 BILLION AIRPLANES, WHILE A PATCH OF PRIMEVAL REDWOODS—NOT TOO BIG FOR A MAN TO WALK THROUGH IN A DAY—WAS CONSIDERED BEYOND ITS MEANS.

The 91,000-acre park at Redwood Creek proposed by the Sierra Club was estimated to cost $150 million, or 75 cents per American. Those opposed to this plan estimated it at double that amount.

Most of the Sierra Club arithmetic was disputed by the lumber companies. Said Arcata Redwood Company: * "Contrary to the impression created by the Sierra Club, redwoods are not disappearing. At least 85% of the original 2,000,000 acres of redwood lands contain redwood stands today. This ad also failed to point out that a significant percentage of this tremendous acreage contained less than 2% redwoods and that only a fraction was in pure stands of the large redwoods such as are now in parks and other reservations."

Arcata maintains that it carries on a forestry program second to none in the forest industry: "All of its lands are immediately reforested with redwood or with other species where these species were growing originally. At least 20% of the original stand is other than redwood."

These arguments are amplified by the Georgia-Pacific Corporation, a larger company than Arcata: *

The truly park-like, outstanding redwoods have already been saved. There are over 140,000 acres of redwood forest land in the 28 state parks and other protected status. More significant, these state parks contain almost all the truly park-like groves still in existence—over 1,500,000 trees,

* Personal letter to author.

enough big ones to make a solid row from New York to San Francisco. In fact, experts believe these state parks contain roughly one-fourth of the superlative cathedral-like groves there ever were.

There never were 2,000,000 acres of such cathedral-like groves as implied, Georgia-Pacific insisted. Outside the state parks most of the so-called redwood forests are ordinary commercial timber lands, with fir, hemlock and other species intermixed with redwood, located generally on steep slopes. (A conservationist might well bristle at the term "commercial timber lands," which implies that a tree has no other value than conversion into boards.) The solid stands of giant redwoods occur only on flats and benches along streams, where flood and fire have destroyed competing species and successive layers of silt nourished the redwoods to their extraordinary size. And they claimed that there are still 1,900,000 acres of forests containing varying percentages of redwoods, with mature trees and trees in varying stages of growth under sound, sustained-yield management.

To these arguments the Sierra Club replied (*Sierra Club Bulletin,* March 1967):

> This statement [that one-fourth of the cathedral-like groves were already saved] has never been documented, let alone proved. Some foresters believe that perhaps 5 per cent of the original 2,000,000 acres of redwoods grew on river flats when logging first began. This would amount to 100,000 acres. But only 3,300 acres of superlative groves have been set aside in California state parks—possibly 3.3 per cent of the original acreage. Furthermore, much of what has been saved is in a series of small, isolated groves which, while lovely to look at briefly—as in a museum—or to drive through, do not offer the possibility of really experiencing a redwood forest.

As for decrying the "slope-type lands," the Sierra Club said, "over 90 per cent of the virgin groves presently in state parks should be called 'slope-type.' Slope-type groves are often exceedingly beautiful. . . . Redwood slopes are also critically important for protecting watersheds and scenic panoramas."

As for "sound, sustained-yield management," said the Sierra Club, massive erosion takes place under present logging practices, and sustained-yield logging may not be practical in many areas because of soil loss. The industry is reseeding to Douglas fir, spruce, and

Monterey pine after cutting its old-growth redwood. Even where redwoods are reseeded, there is a major difference between old-growth, or primeval redwoods, and the second-growth trees. Matchstick forests and millions of stumps are hardly comparable to the forests they replace.

In its own publicity, Georgia-Pacific quoted Newton B. Drury, former director of the National Park Service, later director of the Save-the-Redwoods League, as having said twenty years before: "The best and most representative of these forests are now in California state parks. They are being carefully and intelligently preserved."

If this was the feeling twenty years ago, it was not the feeling of the Save-the-Redwoods League in 1967, or in 1969, a year after the Redwood National Park was created. This organization continues to issue appeal after appeal for funds to buy additional redwood land, sounding the desperate note that time is running out.

Since 1918 the league has raised almost $15 million by public subscription. Every dollar thus raised has been matched by Federal or California funds, and with this money the league has quietly bought redwood land, which it has deeded to the state for addition to its parks. The money has purchased more than 102,000 acres, of which more than 50,000 acres contain virgin redwoods. These acres are distributed among the existing 28 state parks. The estimated value of this land would be at least $250 million if purchased today.

The league does not today believe that the best redwoods are all safely in state parks. Its program calls for buying at least 50,000 acres more of privately owned land, based on the premise that only by enlarging the present parks can their ecological integrity be preserved. For example, there is little point in preserving a stand of redwoods if on the slope above them the trees can be lumbered off without restriction. The resulting floods will wash out the protected area—a situation that has already occurred more than once.

The Ford Foundation has pledged a million dollars to the league on condition that two dollars be raised by the league for each dollar contributed by the foundation.

One of the more curious arguments advanced against a redwood park was that protection may cause them to disappear. In a *Reader's Digest* article of December 1966 Dr. Edward C. Stone, professor of forestry at Berkeley, is quoted as saying that fire and flood are natural events in redwood country; that the trees are dependent upon fire to keep out competing growth, such as Douglas fir and oak; that they are also dependent upon floods to bring in silt, which carries needed nutrients; and that, therefore, a park in which fire and flood are controlled might cause the redwoods to disappear in 100 to 200 years.

It is true that fire—to which the big trees are relatively immune—plays a part in keeping undergrowth down in redwood groves. And normal flooding brings in silt, which nourishes the shallow-rooted giants. But the fear that fire and flood will be eliminated in parks appears to be one that conservationists could endure with equanimity. Apart from the fact that we have never been able to eliminate either one, gaining 100 or 200 years would seem like a genuine boon. For at the present rate of cutting—32 million board feet a year—we will see the end of all the old trees not protected in 15 or 20 years. At which time the arguments of the lumber industry that the bread is being snatched from their mouths will be largely academic anyway.

However, one conservation group, the National Wildlife Federation, took a more sympathetic tone toward the lumber industry during the controversy, adopting the industry view that the best trees were indeed already protected. "Over-emotional statements that the last redwood will soon be destroyed," said Thomas L. Kimball, executive director, "are pure bunk."

But writing in *American Forests* (March 1967), Mr. Kimball describes the result of a flood caused by lumbering the watershed above a protected redwood stand. This flood in 1955 destroyed 300 magnificent primeval redwoods along Bull Creek. The destruction, wrote Mr. Kimball, is even difficult to describe. "Great chasms where none existed before, undercut banks along streams, choked with mud and debris, landslides which uprooted and hurled hundreds of the largest and most beautiful of the mammoth specimen trees into the river like so many matchsticks."

Conservationists agree that ecological integrity is indispensable, that a park must be a complete ecological unit, with its own watershed boundaries. Otherwise it is at the mercy of erosive conditions that spell doom. Victims of such poor planning are the six rivers of this California region: the Eel, Klamath, Trinity, Mad, and Van Duzen rivers. Bad logging practices on their watersheds have caused erosion and sedimentation, filled up natural holes in the stream beds, burying fish eggs and smothering insects upon which fish feed. Pools fill with mud, logging wastes choke the riffles and deplete the oxygen, exchanging it for toxins that kill fish.

As an example of what can happen, heavy rains in December 1964 dumped 20 to 30 inches of water into the valleys of the Klamath, Eel and Smith rivers. The hills, stripped of their forest cover, became raceways for swirling water and landslides of mud. Downstream, towns were inundated in water and silt. Some towns were completely washed away. Damage was estimated at half a billion dollars. Millions of tons of topsoil were carried from upstream and deposited in lower areas in layers up to three feet thick. Several people were drowned, leaving us with an interesting moral dilemma: Who is guilty of their deaths?

Flooding is undoubtedly part of the natural cycle in redwood country, but flooding on a moderate scale. A forested hill is a giant sponge, soaking up water and releasing it gradually. The silt that comes down is deposited slowly, thin layer on layer, and on these rich deposits the coastal redwood grows to its most magnificent size.

Lumber companies in the redwood country do not believe in selective cutting, taking a tree here and a tree there. The giant redwoods, for all their size, are shallow-rooted, and the timber companies say that selective cutting opens up spaces and lets in the wind, which uproots many trees. It is more practical, therefore, to practice clear-cutting, which is to say taking out everything, leaving the hills stripped, presumably for replanting later.

The Sierra Club has some unbelievable pictures of clear-cut areas where the rains have come before the replanting. With the cover destroyed, the water roars down, cutting huge gullies and washing out countless tons of soil. What should have been a "normal" flood becomes an outsize flood, taking with it the humus and topsoil that would have made replanting possible.

During the floods of December 1964, writes Russell Butcher, the Audubon Society's conservation specialist, "the upper few miles of Prairie Creek, protected within the Prairie Creek Redwoods State Park, north of Eureka, sustained practically no damage from high water."

In contrast, according to park supervisor Glen J. Jones, "the lower watershed below the park ran mud until the first of March."

Fighting to keep the Redwood Creek area in the national park proposal, the Sierra Club argued that the 90,000 acres sought was hardly overbuying. It would have increased the percentage of redwood forest land preserved in parks from 5 percent to 8.8 percent. More than 91 percent would still remain available for commercial purposes. "If the industry cannot survive on this preponderant share," said the Sierra Club, "this additional fraction is not going to save it."

Comparing the two areas in question, the Sierra Club added, "A park in the Mill Creek area might cost $56 million, while the park in Redwood Creek is priced at $140 million. However, the money invested in Redwood Creek would save 30,000 acres of virgin redwoods, while the money invested at Mill Creek would save only 7,800 acres of redwoods. Thus, almost twice as many redwoods would be saved by each dollar invested at Redwood Creek. More of the money at Mill Creek goes for acquiring sawmills and private homes located within the boundaries of the project."

Russell Butcher analyzed the many plans in the *Audubon* Magazine for July–August 1965. He agreed that the Sierra Club proposal provided adequate scenery and watershed protection. Half of the 90,000 acres would be virgin redwoods and encompassed a stand of the world's tallest specimens. It would also include the great ocean beach at Gold Bluffs, with its herds of vanishing Roosevelt elk, and the rare groves of huge trees in Prairie Creek Park. The Redwood Creek Valley, he agreed, contains the largest unbroken stretch of primeval redwood forest left.

The Mill Creek area, originally proposed by the Save-the-Redwoods League, and the core of the administration plan, came in for criticism.

The primary problem at Mill Creek [wrote Mr. Butcher] is watershed protection. As at Humboldt State Park, where upstream logging stripped the hills, Jedediah Smith State Park is faced with increasing vulnerability to floods. Only last year a new lumber mill was constructed in the heart of Mill Creek Valley immediately upstream from the park boundary, and the remaining privately owned virgin forest is vanishing under full-scale logging.

In addition to the fact that more than half of Mill Creek Valley is cutover land, other serious questions have been raised concerning the suggestion that this area be made a national park. Although it contains many giant redwoods—one tree has a diameter of 20 feet—the forest at this northern end of the redwood belt contains more trees of other species than do redwood areas further south.

The implication was clear that if this area was to be the basis of the new park, the American public was going to be shortchanged again.

In the end Public Law 9-545, the Redwood Park Act of 1968, turned out to be exactly the kind of patchwork crazy-quilt affair that the Sierra Club had feared, with so many "ifs" and "buts" about it that on August 25, 1969, *The New York Times* was constrained to report: "Nearly a year after its dedication, the National Redwood Park in California still exists largely in theory. And there is a possibility that it may never encompass more than about half of the 58,000 acres specified in the 1968 law that created the park."

The problem lies in the fact that the act authorized the purchase of bits and pieces of land connecting existing state parks, predicated on the assumption that the state of California would then turn its parks over to the Federal government in order to link everything up in one new and larger park. It hasn't quite turned out that way.

The act authorized Federal expenditures of $92,000,000 to buy private lands. By August of 1969 the Department of the Interior had made two large down payments: $18,200,000 to the Arcata Lumber Company for 11,000 acres, and $4,124,931 to the Georgia-Pacific Corporation for 3,422 acres. The transactions were complicated, involving negotiations for land exchanges with the government. And in late summer of 1970 Arcata filed a claim for $121,585,000, asserting that this was the actual value of its 11,000 acres, compared with the government appraisal of $46,240,000. According to the *Times,* Governor Reagan was supporting the lumber companies in driving a good bargain for their trees.

Reagan is also proving a hard bargainer for the state parks to be included in the new national park. He wants some Federally owned shoreline tracts in exchange, and while these have never been exactly specified, there have been some references made to Fort Ord, owned by the army, and to Camp Pendleton, the Marine base near San Diego. The Marines have already turned over to the state a mile of beach south of San Clemente.

Meantime the latest bulletins of the Save-the-Redwoods League, while acclaiming the establishment of a national park as a triumph for conservation, only point up its inadequacy. Their new maps show large sections verifying Mr. Butcher's conviction that they must be acquired for watershed protection, and the league continues to plead for funds to support new purchases of land. Since the establishment of the theoretical national park, the league has spent $785,000 for one tract of 1,316 acres and $156,000 for another piece of 57 acres.

To protect the Bull Creek watershed the league is seeking $100,000; and to buy a piece of private land of 1,200 acres still inside Humboldt State Park the league is working to raise another $700,000. Truly, as they say, "the task is far from finished."

But there is a small, cheerful note. Georgia-Pacific, the lumber company that fought the idea of a national park with bitter self-righteousness, has made a gracious gesture in deeding two groves of redwoods on the Van Duzen River to the Nature Conservancy. The groves total 390 acres and contain many prize trees. Nature Conservancy has turned the tract over to the state of California.

Cynics may wonder if Georgia-Pacific would have been so gracious had it felt it had lost badly; but that is the function of cynics.

PART III

The Wildlife

CHAPTER

6

Our Vanishing Wildlife

FOR the gourmet, hard times lie ahead.

Caviar is becoming vexingly scarce. Salmon is not as plentiful. Oysters and crabs are neither as succulently large nor as bountiful as they once were. Shrimp are now often grossly contaminated with dieldrin. Tuna and abalone are dwindling and green turtles are on the very edge of extinction.

This gloomy gustatory reckoning in our thesis is not entirely frivolous. While much of mankind's food is cultivated, sea fare is a wildlife resource and still is hunted or netted much as primitive man did, albeit with technical improvements.

Marine animals provide a fifth of the world's protein—much more in a country like Japan: perhaps half again as much as is produced on farms. Because in many countries agriculture cannot supply the needed amounts of protein, "the race to loot the sea," as Paul Ehrlich calls it, is now in full swing.

Japan, Russia and Peru are taking the largest share of fish—almost 40 percent of the total catch. The United States and Norway between them get about 10 percent more.

The fishing competition is sharpening. Russian boats probe the invisible borders of American territories, and occasionally one is boarded by the Coast Guard. Americans, in turn, have been stopped by Peruvian naval vessels, which took 24 United States fishing boats into custody between 1961 and 1969, while 50 more were apprehended by other Latin American navy patrols. Mexican authorities have complained that Japanese fleets have violated their territorial waters. A growing tension over fishing grounds suggests that the 100-mile offshore limit may soon become standard in place of the 12-mile limit.

With fish catches more than doubling between 1953 and 1967, an ancient drama is being repeated. As the bigger and choicer fish become scarcer, commercial fishermen take smaller and smaller ones. This not only deprives the bigger fish of their food but cuts down the reproductive potential of the entire species.

Reproduction, meantime, is being further hampered by the increasing pollution of the seas. The Norwegian explorer Thor Heyerdahl found heavy pollution not merely around harbors but far out to sea. Heyerdahl tried to sail from Morocco to America on a raft made of bundles of papyrus reeds to test a theory that the Egyptians might have made that voyage. His raft became waterlogged after 52 days and 2,700 miles and had to be abandoned, but it provided a unique opportunity for observation. At an average speed of only two knots or so, with the raft platform barely more than a foot above the surface, they could examine the water closely.

In mid-ocean the crew sailed through polluted water, dirty and grayish-green instead of blue, as though "amidst the outlet of city sewers." They also saw "brownish to pitch-black lumps of tarlike or asphaltlike material of the size of fine gravel and dispersed at irregular intervals on and slightly below the surface."

This corresponds to other scientific reports that the ocean is now almost uniformly contaminated with globules of oil or tarlike material which are being swallowed by fish and are quite toxic.

But even before pollution had become a serious factor in fish ecology, indiscriminate catches had raised the specter of extermination for more than one sea creature. Whalers, for example, still go to sea, although the industry grows visibly more desperate as the great sea mammals decline.

In the past 50 years whalers have killed more than 2,000,000 whales, bringing the prized blue whale to the verge of extinction. And here is a very revealing statistic: in 1933 whale hunters brought in 28,907 whales, from which were extracted 2,606,201 barrels of whale oil. Thirty years later weaponry had improved and in 1966 they harpooned 57,891 whales—about twice as many. Did they then get twice as much oil? They did not. They got 1,546,904 barrels, only 60 percent as much as before.

Why? Because the biggest and finest whales were gone. They were taking younger whales and smaller species.

The International Whaling Commission reported in 1963 that blue and humpbacked whales were in serious danger of becoming extinct. It recommended a total ban on these two species and a strict quota on fin whales, also reaching the danger mark. That year whalers killed 13,780 fin whales, about 35 percent of the estimated total population and *three times* the sustainable number.

The next year the commission again recommended strict quotas, which were overruled and arbitrarily doubled by the four major whaling countries, Japan, Holland, Norway and Russia.

Then a curious thing happened. The whalers didn't achieve their own quotas, and Holland went out of the business and sold its ships to Japan.

In 1967 a total of 52,046 whales were killed, half of them sperm whales, which the commission had also tried unsuccessfully to protect. The Japanese, hunting protein sources, also killed 20,000 porpoises. And the next year Norway gave up and dropped out of the whaling business.

There are few more concrete examples of an industry cutting its own throat. It is now reported that Russian whalers, unable to find enough of the cetaceans, are harvesting krill, the little antarctic shrimp that furnish the basic food supply for most whales.

Even if one has no ecological, sentimental or any other kind of reason for preserving whales, if one wishes only to catch and kill them, it still makes more sense to leave the krill for the whales to eat and then harvest adult whales than to take their food supply and so eliminate both krill and whales. Apparently even this rudimentary judgment seems to abandon those afflicted by greed or quotas.

Early in 1970 an international battle over salmon erupted. It began on a slightly comical note, although the implications are not humorous.

The Duke of Roxburgh, Scotland's premier baronet, made vigorous vocal complaint that he had hooked only nine Atlantic salmon instead of the hundred or more he was accustomed to haul from the river below his ancestral castle. The British government, backed by Russia, Canada, the United States and several other countries, promptly pointed the finger at Denmark as the spoiler. The Danes, they said, had been catching salmon on the high seas and, since Denmark has no salmon streams of its own, doing nothing about restocking. The Danes at first angrily rejected the accusation but then agreed to talks on saving the salmon.

The problem with salmon is wrapped up in an uncommon adjective: anadromous. An anadromous fish is one that is hatched in a fresh-water stream but emigrates to the ocean to feed and mature. Fully grown—in about four years for the salmon—it returns to the stream of its birth to spawn. This life cycle of the salmon makes it, next to the whale, the most harried creature of our time.

On our side of the ocean the Atlantic salmon spawns in some streams of Maine and Canada. The Connecticut River, once thick with salmon, has seen none for years because of extreme pollution.

In Europe, salmon spawn in the streams of Norway, Scotland and Ireland, about the last places where clean water may be found. Once they spawned all the way to the Mediterranean, but dams and pollution have eradicated them from France, Holland and Portugal.

The young salmon spend the first year or so of their lives in the stream where they hatched and then start for the sea. Once there they vanish. No one trawled for salmon on the high seas because no one knew where they were. The big catches were always made when the great schools of fish returned to spawn in the rivers of their birth.

But in 1964 Danish fishermen discovered a major salmon-feeding area in the Atlantic off the southwest coast of Greenland. Their catches increased from 36 metric tons in 1965 to 1,000 tons in 1969.

Coincidentally the catches elsewhere went down. In Canadian rivers it went down 45 percent in 1968. In London only 40 tons of salmon were sold in 1969 as against 105 tons in 1963. In Scotland salmon catches dropped by 25 percent.

The protests of ten countries were carried to the International Commission for Northwest Atlantic Fisheries, where, at a meeting in June 1969, the demand was made to ban salmon fishing on the high seas. Denmark called it unreasonable but agreed to listen to less restrictive proposals.

The following March, at a monthly meeting of Scandinavian ministers, Denmark agreed to limit its catch but not to entertain an outright ban on salmon trawling at sea.

The meeting was held at the Hamlet Hotel, a small inn just outside Kronberg Castle in Elsinore, the locale of Shakespeare's play. It can only be conjectured that *läx*—smoked salmon—was on the luncheon menu, but it is reliably reported that Danish Fisheries Minister A. C. Normann remarked, "When someone said in the British Parliament recently, 'Something is rotten in the state of Denmark,' I guess they were referring to me. We seem to be blamed for everything."

Pollution and disease, said Mr. Normann, were the major causes of the salmon shortage, not Denmark's interception of the fish on the Atlantic feeding grounds. However, the other countries claimed that since Denmark has no salmon streams of its own, the Danes were actually taking the salmon of Canada and Britain, where millions were being spent in restocking.

Most of the salmon caught in the United States are Pacific salmon, and a somewhat different situation exists on our West Coast. Salmon hatched in the Snake and Columbia rivers start downstream to sea past some of the most formidable hurdles ever placed in the way of fish or man, an obstacle course of huge hydroelectric power dams.

The fingerlings have a choice: they can go through the generating turbines, which might chop them up, or over the spillway, where the fall might knock them senseless. Whichever choice they make, many are killed at each dam, others injured or so debilitated that they are unable to avoid predators; so a reduced number reaches the coast.

After a period in the estuary to become acclimatized to the salt water, the salmon head into the sea. Most follow the currents north toward Alaska, although some have been found heading south along the California coastline.

Fully grown in about four years, they head back to spawn, guided by some navigational instinct through thousands of miles of ocean, back to the same river, branch, tributary or creek where they began. And from the time they hit the coast again, all the long way up the river—as much as 2,000 miles in the case of the Yukon Chinook—at every foot of the way some predator waits or some barrier looms.

Columbia River salmon once fought their way upstream a thousand miles to the headwaters in Canada. Now the journey ends at the 600-mile mark, barred by the Grand Coulee Dam. They have no way of getting past this high dam since it was built without the fish ladders of more modern structures.

Salmon do not eat once they enter the river and begin the upstream swim, living off their stored fat. Fighting the current for a thousand miles without food is bad enough, but when the way is barred by one high dam after another, even with fish ladders, each is an obstacle that exacts a toll of strength and stamina. Fewer and fewer of the salmon reach the next dam. And so their numbers dwindle.

In the interminable battle between those who would protect wildlife and those with contempt for this sentimentality, the argument is sometimes heard that far from dying out, wildlife is more plentiful than ever. Deer, it is said, are more numerous now in the Northeast than in George Washington's time.

New York State, where deer were extinct in 1915, now has about 400,000. In Westchester County, an exurb of New York City, deer stroll down the streets of villages, boldly browse in backyard vegetable gardens, and generally make themselves at home despite throughways, cars, dogs and hunters.

Ring-necked pheasants, introduced from China, have spread across the country, wild boars brought in from Prussia flourish in the mountains of North Carolina, and mourning doves are said to be plentiful despite the fact that hunters are shooting them at the unbelievable rate of 20,000,000 a year. Raccoons and opossums seem actually to benefit by man's presence and handiwork.

These, however, are hardly the rule. More than 850 vertebrates—and even more species of plants—are on the endangered-species list of the International Union for the Conservation of Nature. Hundreds of animals have already been exterminated down to the last survivor.

How was it possible, asks ornithologist Edward Howe Forbush, for man to kill all the passenger pigeons? In the eighteenth and nineteenth centuries the American passenger pigeon was the most numerous creature on the continent, its numbers literally in the billions.

Alexander Wilson, the founder of American ornithology, described a flight of pigeons in 1806 that he estimated to be more than a mile wide and about 240 miles long and containing more than 2,230,000,000 birds. The flight rolled over his head from one-thirty to four o'clock in the afternoon.

In 1813 John James Audubon left his house at Henderson, Kentucky, to ride to Louisville. A flight of wild pigeons came over and "the light of noonday was obscured as by an eclipse." He traveled the 55 miles from Hardenburgh to Louisville under a vast umbrella of birds streaming overhead and darkening the sun. The flight continued for three days. The people turned out on the Ohio River bank, where the pigeons flew lower over the water, and fired their guns into the flock, bringing down so many thousands that for a week they ate nothing else. Dozens of birds were killed with a single shot.

Wherever the pigeons came to rest they were caught in nets, shot, or killed with clubs and stones. Audubon describes a roosting place near the Green River in Kentucky that was 40 miles long and three miles wide. The birds, he said, came in soon after sundown with a roar of wings "like a gale passing through the rigging of a close-reefed vessel," creating a wind as they came. They alighted in the trees in such numbers that limbs broke off and hundreds of birds went crashing to the ground. The uproar was incredible. Trees as big as two feet in diameter broke down from the weight, and branches of the tallest trees gave way. No one, he says, dared venture into the woods at night because of the danger of falling branches.

The Indians found the pigeons a reliable source of food but made no inroads into their numbers because they took only what they needed and did not organize a commerce in squabs, as the whites did.

Commercial fowlers, using nets, followed the flocks wherever they went, sometimes a thousand miles cross-country. One of the last great slaughters of pigeons occurred in New York State about 1870. The flock had originated in Missouri in April of that year, where most of the spring squabs were taken by the hunters. The flock then migrated to Michigan, but were followed by the hunters, who took the second crop of squabs. The pigeons then moved to New York and nested in the headwaters of the Beaverkill Creek in the Catskill Mountains. The hunters found them again. Tons of the birds were killed, packed in ice and sent to the New York market. New York for many years had been receiving about a hundred barrels a day.

Meantime the forests were being cut down, and with them the feeding and nesting grounds of the pigeons. The seeds of beech, oak, pine and hemlock were the pigeons' staple food, and as the forests went, their subsistence was cut from under them.

After 1878 the pigeons dwindled in North America. Ernest Thompson Seton, the great Canadian naturalist, recorded the last great flight in Manitoba in 1878. By 1893 sightings had become rarer. In 1900 about 50 birds were reported. The last passenger pigeon died in the Cincinnati zoo in 1914.

Audubon did not believe the pigeons could be exterminated, because of their incredible numbers. How could man kill so many billions? But it isn't necessary to kill every individual to exterminate a species. All that is required is to kill a great number of the young each breeding season. Under optimum conditions only a certain portion of each brood survives. Kill enough of the rest and the reproductive blight is out of proportion to the actual numerical toll.

As a very low figure is reached, another natural law seems to come into operation: further reproduction becomes chancy or unlikely. Once the number of individuals falls below some minimum, the future of the species is in very grave doubt.

Biologists are now trying to rear the young of some endangered species in protective custody with a view to reintroducing them into the wild when they are able to fend for themselves. Dr. Paul S. Martin of the University of Arizona has even suggested that we attempt to raise some close relatives of lost species and see how well they fare under modern conditions.

Some of the extinct camel-like animals of North America were able to feed on mesquite and creosote bushes, which modern cattle cannot digest. American farmers, says Dr. Martin, are "hung up" on cattle and keep trying to eliminate the woody plants of the American desert with herbicides in order to replace them with grass, which is unsuited to the climate. It might be a more rewarding experience, he thinks, to replace the cattle with a kind of livestock that can thrive on the native plants.

Since 1965 the Patuxent Wildlife Research Center has been developing a nature bank of rare and endangered species. The goal is to build a kind of Noah's Ark of animals that can be reintroduced where they have been eliminated.

Biologist Ray C. Erickson of the Patuxent Center visualizes a number of substations for different climates—one in the Gulf states for tropical and subtropical species, one in the southwestern states for desert animals, and one in Hawaii for oceanic and tropical-island species.

The animals we know today were not the original inhabitants of North America. A million years ago, so runs the current theory,

Alaska and Siberia were joined by a solid land bridge, and over this bridge came the ancestors of our familiar animals: wolf, bear, fox, moose, elk, bison, and a little later, geologically speaking, man the hunter.

The men who reached Alaska during this Pleistocene period found animals now long vanished—the Titanotylopus, a giant camel; Eremotherium, a huge ground sloth; the woolly mammoth; the big-horned bison, and various forms of tapir, horse and camel.

The horse actually originated in North America, spread to Asia and became extinct in America. It then crossed back into America from Siberia and again became extinct on this continent until reintroduced by the Spanish explorers in the sixteenth century.

Obviously weak in comparison with the saber-toothed tiger and the great mammoths and mastodons, man successfully ran down and killed these formidable creatures. He did it by teamwork with other men and by creating the simplest of weapons—the spear. With this weapon primitive man began a career of extermination that has never flagged and that, with more efficient tools, operates even more ingeniously today.

Is it unthinkable that these early hunters, with flint-tipped spears, lacking even the most primitive bows and arrows, could have contributed not merely to the killing but the actual extermination of the mammoth and the big-horned bison?

Primitive man was no conservationist. A favorite method of hunting was to stampede an entire herd over a cliff, where hundreds of animals were killed by the fall. The hunters took what meat they could carry, and the rest was left to the scavengers.

Another technique was for bands of hunters to set grass fires to drive the big animals into a bog, where they mired down and could then be surrounded and killed at leisure.

This kind of slaughter was a significant change from the normal hunting pattern of predators, who were generally unable to kill more than the single animal they could eat. The killing of the herds reinforced the effects of the changing climate as the ice receded, and paved the way for the emergence of the modern animals—bison, deer and antelope. This great shift in animal populations was comparable in scale to the ending of the dinosaur age many millions of years earlier.

Extermination of species, then, is nothing new; it has happened many times. But in North America it seems that man has played a major part in the process, because his own species had already become numerous enough to swing the balance begun by the changes in climate. In *Man's Rise to Civilization* * Peter Farb writes:

> Although the ice-age hunters were relatively few, the inroads they made into mammal populations were out of all proportion to their numbers. Early man concentrated around waterholes and streams; as the land became more arid, the huge beasts had to go to those watering places where man himself was lying in wait. There are piles of mammal bones at Folsom and Plano sites, and they are not accidental, for a graveyard of mammals no more existed than did the mythical one of the African elephant. Rather, the bones tell a story of repeated slaughter by Paleo-Indian hunters along the watercourses.

Similar exterminations did not occur in South America in spite of climatic changes because the human population had not become large enough to tip the balance.

Today it is large enough everywhere, and today the business of extermination goes forward more briskly because man is competing actively for space with the remaining wildlife and because a myopic technology is making the planet hazardous for all forms of life.

There are currently about 89 species of animals and birds on the Department of the Interior's endangered-species list. These are included in the worldwide number of 850, for in other places the same pulverization of wildlife is going on.

In Australia the koala bear, prototype of the teddy bear, was slaughtered by the millions. The golden marmoset is vanishing from Brazil. The orangutan is gone except for a remnant in Sumatra and Borneo. The Philippine eagle is down to fewer than 100 birds. The fate of the Indian tiger, the African lion, the elephant, the rhino and hippo are in serious question. The kangaroo, unofficial emblem of Australia, may be doomed. It is facing the same fate as the American bison and for the same reason—it competes with livestock for grass.

A single Australian sheepman claims to have killed 20,000 kangaroos in four years. A professional kangaroo hunter estimates his own kill at 50,000 in five years. A government publication, *Wildlife Service*, says the kangaroo meat and skin industry has absorbed at least 1,500,000 animals a year, mostly from New South Wales and

* E. P. Dutton, 1968.

Queensland. Two million pounds of kangaroo meat for pet food, worth $400,000, were shipped abroad in 1969, while the sale of kangaroo skins brought in nearly $1,500,000—half of this from the United States. It is all quite reminiscent of similar episodes in our own country.

In Bolivia and Peru it is a criminal offense to kill the vanishing vicuña or to offer its wool for sale. But the world's most prized and expensive wool is still available in certain markets of the world, where it brings prices as high as $240 a yard. The governments of Peru and Bolivia have outlawed exports of vicuña wool or hides for ten years, and the United States has closed the doors to their importation, but so long as other markets exist, the poachers have not stopped killing these shy animals of the high Andes.

Felipe Benavides, a Peruvian contractor who loses a good deal of time from his business to defend the vicuña, believes the only way to save these animals is by the concerted efforts of governments to close all markets for wool and hides.

The same idea began to gather momentum in the United States in 1970. Congressman Henry S. Reuss of Wisconsin had paved the way by asking the United Nations to organize a world movement for the protection of endangered species, citing the high risk to the leopard because of a fad for leopard coats.

In February 1970 Governor Nelson Rockefeller of New York asked the state legislature for laws making it a misdemeanor to possess or sell the hide or any other part of an endangered species. Two laws were drafted, one of which, the Mason bill, was the first to name specific animals, although it trailed by more than a month an action by Mayor John Lindsay of New York City in banning the sale of alligators or alligator products in the city.

The Mason bill listed 14 mammals, 47 birds and 21 fish. It covered leopards, tigers, cheetahs, alligators, vicuñas, polar bears, cougars, jaguars and ocelots, to name only the better-known. Among the more exotic species were the Indiana bat, the Hawaiian dark-rumped petrel, the o-o-a-a (another Hawaiian bird), the Texas blind salamander and the short-nosed sturgeon.

At the Albany hearings on the Mason bill Edward C. Classen, representing the fur industry, told members of the agriculture committee that passage of the law would make the New York furrier an extinct species. Some animals singled out for protection, such as ocelots and margays, he said, were actually quite plentiful. The future of the fur industry, he maintained, depended on a continuing harvest and supply of pelts, hence the preservation of endangered species was just as much a matter of concern to the fur industry.

He went on to say that a bill barring the sale of pelts of such animals as leopards, tigers, cheetahs, polar bears, cougars, jaguars, ocelots, margays and alligators would drive buyers from New York and cause a further decline in the New York fur industry. Apparently there was nothing contradictory to him in the fact that most, if not all, of these animals are on the endangered-species list.

The industry's concern over the New York law had meanwhile spread to Washington, where Secretary of the Interior Hickel was considering possible revisions of the list, with subsequent revisions of Public Law 91-135 forbidding importation of such species to the United States except for research. The revisions were expected to include gorillas, snow leopards and cheetahs.

Gorillas are not much in demand in the fur market, nor are snow leopards, cheetahs or polar bears. But the pain being experienced by the furriers was over a larger issue—many women's new sensitivity toward wearing furs as a result of the activity of conservation and humane societies.

Manufacturers in Manhattan's fur district had been picketed, as had department stores selling furs and alligator-skin products. Letters had been written to newspapers decrying the slaughter of animals for their pelts, and groups of people had signed pledges not to buy coats or articles made from the skins of wild animals.

The prospect of such a movement's gaining strength evoked furrier alarm. Alfred B. Cohen, chairman of the Retail Fur Council said (*New York Times,* March 1, 1970), "We are not opposed to conservation, but we are opposed to conservationists who do not take the trouble to obtain the facts. There is a new Federal Law, Public Law 91-135, to which we all must abide and which we cooperated with the Department of Interior in formulating. The conservationists too know that we have a new law, but that does not prevent them from engaging in their inciting tactics."

William Fitzgerald, president of the New York Fur Auction and the Council of Fur Organizations, said picketing of stores was "very disturbing. The idea that everything that the consumer buys with fur on it represents an animal species that is being slaughtered gives a false impression."

It does?

The executive vice-president of the Associated Fur Manufacturers' Association quoted the Red Data Book, prepared by the International Union for Conservation of Nature and Natural Resources, a Swiss organization, to show "that very few species are actually threatened with extinction. Of the few species threatened with extinction, only two or three are used for fur purposes—the majority are killed in defense of humans and domestic animals."

The Red Data Book is now considered significantly out of date; the business of extinction marches on more briskly than anticipated.

One animal definitely threatened with extinction, but less by furriers than by government action, is the American cougar—also called puma, panther and mountain lion. The cougar is a rather timid beast for all its size and armament, and has never been known to attack a man. Its prey are deer and related ungulates. The decline of the cougar (and wolf) is one of the reasons there has been a boom in deer populations in some areas.

In 1903, when President Theodore Roosevelt visited Yellowstone National Park, he remarked that the cougar was the only foe of the elk. "The cougars," he said, "were preying on nothing but elk in the Yellowstone Valley. As the elk were evidently too numerous for their feed, I do not think the cougars were doing any damage."

Despite this statement, Roosevelt did not interfere with government hunters, who, over the next quarter of a century, systematically exterminated both cougars and wolves from the Yellowstone.

The elk responded by mushrooming in number to such an extent that during the winter of 1961 and 1962 government hunters shot 5,000 of the animals. A few years later, following an order to shoot another 600 elk, protests from Montana and Wyoming congressmen and state officials as well as conservation groups stopped the killing after 200 had been shot.

In 1965 Victor H. Cahalane, assistant director of the New York State Museum, author of *Mammals of North America,* did a survey for the New York Zoological Society on the cougar, the grizzly bear and the wolf in North America. His report showed that the cougar had been exterminated from the eastern United States except for an estimated 100 to 300 "pale panthers" in Florida and perhaps 25 cougars in Nova Scotia and the Gaspé Peninsula—although some wildlife authorities doubt the latter. He estimated that 4,000 to 6,500 were left in the mountain states of the West.

The once-prevalent bounties on cougars have generally been discontinued, but western sheepmen, still paranoiac about lions, have kept the cougar in the "noxious animal" category. Consequently there is no closed season and cougars may be hunted at any time. In 1963, the last year for which figures are available, 294 cougars were killed by "control" operations supervised by the Fish and Wildlife Service. This does not include those killed by hunters, stockmen or state agencies.

"The future of the cougar," says biologist Dr. Clarence Cottam, Director of the Welder Wildlife Foundation, "is not bright."

Nor is there much light in the future of the American alligator in spite of recent efforts to save him.

The largest tannery of reptile skins in the country, Peter Baran and Sons, Inc., of Harrison, New Jersey, reported business down 50 percent for 1969 over 1968. And it was expected to go lower, said Peter J. Clancy, president of the company. "I think people have become too emotional," he said. "You have to consider the health of the business and the number of people involved."

Mr. Clancy may be overlooking the fact that it is not emotion but a lack of alligators that has caused his particular depression. Some years ago Baran and Sons had 40 employees and handled 200,000 reptile skins a year. In 1970 the firm had six employees treating fewer than 12,000 skins a year.

The American alligator, *Alligator mississipiensis,* is not as formidable a beast as the crocodile, but a good-sized male will stretch 12 feet and weigh about 500 pounds. His powerful jaws can readily crush a big turtle, shell and all, and his stomach acids can digest it. Most alligators are concentrated in Florida and Louisiana, although some appear as far north as North Carolina.

About the middle of the nineteenth century a tanning process for reptile skins was developed, and the rush for alligator bags, shoes, belts and wallets was on. The alligator hunters descended on the swamps with guns, spears, hatchets and machetes. Boats were chartered by sportsmen and commercial hunters, and they returned from the marsh laden with dead adults and live baby alligators to be sold as pets.

There is, in fact, a recurrent fable that many of these pets are now alive and well in the sewers of New York, full grown and living high on rats. They are supposed to have become established in the sewers after being flushed down the toilet by panicky mothers who weren't happy about their children receiving baby alligators from doting uncles in Florida. The warmth of bacterial fermentation in the sewers is said to be sufficient to keep these subtropical reptiles quite comfortable.

The alligator-skin business flourished well into the present century. One estimate holds that by the year 1900 more than 3,000,000 alligators were killed in Florida alone.

Observing the rapid destruction of a natural resource, Florida imposed a ban in 1944 on hunting alligators during the breeding season. The ban was lifted in 1950, with a restriction that only alligators over eight feet long were to be taken. The kill was 18,735. It was considered too high, and a closed season was imposed. But closed seasons or any seasons meant little to poachers, who went right on killing about 50,000 alligators and reaping an illegal reward of about $1,000,000 a year.

Poaching thrives, first, because it is profitable and, second, because enforcement is next to impossible in the vast reaches of the Everglades.

A two-man team can still make $500 a night and faces little danger of being caught. If the men should be caught they may be fined, the maximum penalty being $1,000 or a year in jail. In actuality the heaviest fine ever levied was $750 and the longest jail sentence six months.

From 1961 to 1968 the Law Enforcement Division of the Florida Game and Fresh Water Fish Commission made 597 arrests. Of these, 331 suspects were convicted and 91 acquitted, with the rest pending. Fines totaled $26,125, and total jail sentences about three years. Wardens confiscated 3,686 hides, which were auctioned off for $47,586. And the frustrated wardens complained that the poachers were back in business the next night.

The Florida Audubon Society asked for a tightening of the law and had its answer from the legislature in April of 1970. Saying it was foolish to give alligators more protection than humans, the

House killed a provision for a mandatory 10-day jail sentence for alligator poachers, continuing it as a misdemeanor instead of a felony.

If this seems like a non sequitur, it is. Two state legislators, Representative Gerald Lewis and Representative Jerome Pratt, remarked that the Florida legislature has consistently rejected bills requiring a mandatory jail sentence on a first conviction for drunken driving, even though that offense endangers human life. And Representative Edmund Whitson added that the entire purpose of the bill was to take the profit out of poaching by notifying poachers to expect a jail sentence instead of a small fine.

What good is an alligator? Like the cougar or the wolf, the alligator is a necessary predator in the ecological chain, keeping turtle and rough-fish populations stable, performing essential scavenger duty on carrion, and helping to stabilize the water supply by digging out 'gator holes in which fish collect during the dry seasons. Alligators also provide protection to bird rookeries from marauding raccoons. Waterfowl—ducks, coots, the big wading birds—can generally keep clear of the alligators without much trouble. It is common in the Everglades to see ducks or coots swimming near alligators without concern.

The big alligators are rare now, and even poachers, killing smaller and smaller ones, admit that the reptiles are down to about one-tenth their former numbers. Yet the poachers' attitude is frustratingly human. It isn't "let's save them now," but rather "let's get the rest of them while they last."

A reptilian relative of the alligator, the sea turtle, is in even worse trouble. Like the salmon, the marine turtles are migratory. They come ashore in waves every two or three years to lay their eggs in the sand, and this event is awaited by such a variety of pillagers—animal, bird and human—that fewer and fewer baby turtles make it back to the protection of the sea.

Turtles are hunted for a number of end products: eggs, shell for expensive tortoiseshell artifacts, turtle meat, turtle oil for cosmetics, and turtle leather. The green turtle is near extinction because of a dubious distinction—it alone yields calipee, the cartilage that joins the bones of the belly shell, an essential ingredient in green turtle soup. From a 350-pound green turtle the hunter may cut out three or four pounds of calipee. In the early sixties one New York City firm used 5,000 Caribbean green turtles a year to make 600,000 quarts of soup.

The turtle harvest is reminiscent of the great passenger pigeon slaughter. Tens of millions of turtle eggs are plundered, thousands of turtles killed in the Caribbean area for meat, hides, shell or calipee.

There are only sporadic or uncoordinated efforts at control of the turtle traffic, and even within this loose protection there is widespread poaching. For the green turtle the obvious protection of halting the importation of calipee is yet to come. One attempt to reintroduce green turtles into areas from which they have been exterminated has not yet been proved effective.

For many people the term "endangered species" has become synonymous with the whooping crane, a bird millions of people know by name if not by sight. There are only about 50 of these birds in the world now, but they have received so much publicity that heroic efforts were set in motion to save them.

Of course there is the backlash. As Vice-President Spiro Agnew managed somehow to equate people who would save the Florida alligator with Communist sympathizers, there are those who found themselves irritated by the whooping crane. Glenn Kimble, director of air and water resources of the Union Camp paper mill in Savannah, Georgia, one of the state's larger polluters, displayed annoyance in an interview with a *New York Times* reporter on May 4, 1970.

"People get extremely emotional about losing a species," he is quoted as saying. "But animals have been dying out every year clear back to the dinosaurs, and in most cases man had nothing to do with it. For that matter it probably won't hurt mankind a whole hell of a lot in the long run if the whooping crane doesn't quite make it."

Perhaps so. The loss of the passenger pigeon didn't seem to hurt mankind a great deal. Yet somehow here we are in the seventies, wondering if *mankind* is going to make it. And the Kimbles of the world never do see the connection.

The *planned* extermination of species by a few men because of their special interests, says biologist Roger Caras, is something new and frightening. If it is carried to its logical end, given different men, different interests, different points in time, every living thing will one day be eliminated. A case can be made against any plant or animal.

The whooping crane is a handsome bird, five feet tall, with a seven-foot wingspread. It makes a migratory flight of 2,600 miles each year from the Aransas Refuge in Texas to its summer breeding ground in northern Canada.

Such big birds were a natural target for hunters, who shot them even though they would never dream of eating one. The birds lay only two eggs a year and generally only one chick survives. By 1938 there were only 25 birds left.

The Department of the Interior, under Secretary Udall, thought it worth an effort to save these threatened birds and in 1967 organized Operation Whooping Crane.

In a joint Canadian-United States operation, biologists flew into the Canadian wilderness and deftly began robbing whooper nests. They knew if they carefully removed one egg from a nest the adult birds would not abandon the remaining egg.

The rifled eggs were flown to the Patuxent Research Center and hatched artificially. Out of six eggs taken in 1967 four birds survived. In 1968 seven survived out of ten eggs taken. An attempt to mate captive birds was also successful, and it was learned that the mature birds will breed in captivity and raise their young. Thus the whooper population came back from its low point of 25 to something over 50. But even under protected status a wild bird is occasionally shot, and the species is still right on the danger line, that line of minimum population that makes survival very much a question mark.

An even easier target than the whooping crane is the California condor, our largest vulture, with a 10-foot wingspread. The condor, little changed from prehistoric times, has been called an ice-age relic, a living fossil. Says Dr. S. Dillon Ripley of the Smithsonian Institution:

> The condor represents among the warm-blooded vertebrates, birds and mammals, one of the very few remaining natural genetic reservoirs, unchanged since Pleistocene times a million years ago. It should be preserved as a biological resource.
>
> In the future, scientists may well be put to it to reconstruct genetic combinations which exist today in such pure strains from the past. A whole biological story remains to be told here, waiting for the ever-increasing skills of scientists to interpret. What a catastrophe if the story

were to be destroyed before its meaning could be unraveled, the code interrupted before it is lost.

Condors were once numerous in California and east to New Mexico, possibly even to Florida. Condor eggs became collectors' items, and since the birds lay only one egg every two years, their numbers declined rapidly.

California miners killed them for their large quills, which they used as containers for gold dust. Being scavengers, condors were also killed by poisoned carcasses left for coyotes. By 1966 the California condor was down to 40 birds.

Legally they are now protected. But the biggest threat to the "forty dirty birds," as someone contemptuously referred to them, is still ahead. They are extremely shy, and the sight of a man is likely to drive them from their range. They now nest only in remote wilderness areas, such as the Sespe Sanctuary in California. Such areas are frequently the target of developers, who consider both birds and conservationists interfering nuisances.

The Army Corps of Engineers has been interested in building a high dam on Sespe Creek and furthering the development of this area into a recreational project to accommodate 2,000,000 visitor days a year. The project is visualized with the usual motor camp sites, picnic areas, riding and hiking trails and marinas. It has been stalled for now, but is far from dead. Should it come up again and win, the condors will have no further place to retreat.

Birds of prey generally are high on the endangered list, in the United States as elsewhere in the world. For years biologists have pleaded the case of the hawks, pointing out that their food consists essentially of mice and other rodents and that they are an integral part of the food chain. And for years hawks have been shot by farmers as chicken thieves or by hunters just looking for a moving target.

In its colossal ignorance the state of Pennsylvania in 1885 passed a law known as the Scalp Act, which paid bounties on birds of prey. It led to a program of organized hawk shooting. In the first year and a half of this law's operation, bounties were paid on 125,000 hawks and owls in Pennsylvania.

Hunters discovered that the migratory route of hawks in the East took them past a fifteen-hundred-foot ridge in the Pennsylvania mountains. On a day in late October or early November, the peak of the fall migration, hundreds of hawks swept by this rocky outcropping. And hundreds of hunters, sometimes as many as 200 on a weekend, gathered there too, just to blast the birds out of the sky. Ornithologist Richard H. Pough, who watched the shambles in 1932, made a count that estimated 5,000 birds killed there every year.

About 1934 Mrs. C. N. Edge of New York bought Hawk Mountain and transformed it into Hawk Mountain Sanctuary with the aid of Maurice Broun, who became its first curator. This was done over the incredulous resentment of the local populace, irritated at the intrusion of out-of-state foreigners—and crackpots to boot, as anyone must be who wanted to save hawks.

The hostility has rather died down now, and the local people are quite proud of the sanctuary and the fact that more than 25,000 people a year visit Hawk Mountain in the fall for the incomparable sight of the great hawks—the world's most magnificent fliers—as they sail by, almost close enough to touch.

Organized hawk shoots were commonplace in many states. One report describes the killing of 4,000 broad-winged hawks in two days in Minnesota. In the Midwest, where trees may be scarce and a fence post makes a convenient resting-place for hawks, the bird is an irresistible target for gunners. Roland C. Clement, Audubon biologist, describes a report (*Audubon* Magazine, January–February 1965) of 200 dead Swainson's hawks found beneath 200 telephone poles along one short stretch of highway in Kansas. In many midwestern states barbed-wire fences are adorned with the dead bodies of hawks, strung up in some kind of symbolic sacrificial rite.

In Kenya, East Africa, says Mr. Clement, the Augur buzzard, a bird much like our red-tailed hawk, also sits on every telephone pole along the highway. The natives, although illiterate, never kill these birds. "The red-blooded American boys who did this wanton killing," he says, "would be incensed if I rated them below the tribal Africans of Kenya in intelligence or civilized status. The real difference, however, is the guns in the hands of our outdoor illiterates."

By now at least 19 states have passed laws to protect birds of prey, and another 26 states have partial protection. But already the threat is less from the gun than from pesticide residues, which have resulted in infertile eggs or eggshells so thin they collapse when the mother bird attempts to incubate them.

Everywhere in the world the population of eagles, ospreys and other hawks has plummeted. In Israel, for example, it is reported from the Department of Zoology at Tel Aviv University that intensive use of pesticides has killed 90 percent of the raptorial birds.

Fish-eating birds get a staggering dose of DDT, and, as already noted, the pelicans are fast disappearing. Louisiana's 50,000 brown pelicans are gone. Only a few are left along the Texas Gulf Coast. Audubon biologists are counting and studying Florida's pelicans to watch trends developing there.

In California, signs of disaster—broken eggs—are unmistakable. Dr. Robert Risebrough, of the University of California, found only 12 intact eggs out of 298 pelican nests in one colony. Asked how many young pelicans were raised in California in 1969, Dr. Risebrough said, "At most, five."

Obviously, California pelicans are headed for extinction unless the DDT levels in the sea drop quickly—which they won't.

Horses are not ordinarily thought of as wild animals. But there are wild horses in the West and they are an endangered species. In 1924 Will C. Barnes of the U.S. Forest Service wrote: "With every man's hand against them, these wild horses will eventually be exterminated. In the meantime, any red-blooded man thirsting for adventure, excitement, and some Wild West riding can get plenty of it chasing these unwelcome residents of the western ranges. There is no closed season on them at any time in the whole year, for they are classed with wolves and coyotes as predatory animals, marked for slaughter."

Like Australian kangaroos, the wild horse competes with cattle and sheep for the sparse grass. Stockmen therefore shoot them whenever they can or round them up periodically and sell them to dog-food canners. The Bureau of Land Management estimates there are about 17,000 horses left in the western states.

Once the wild mustangs ran in great bands. They were not the magnificent, gleaming stallions portrayed in unrealistic Western movies—they were half-starved, runty broomtails, desired by nobody except perhaps a cowboy afoot who needed a mount no matter how sorry.

In 1898 the market value of a wild horse was about 24 cents. With the outbreak of World War I and a sudden demand for horses, the price went to $40 a head. Thousands were rounded up and sold, then shipped east under such terrible conditions that the mortality was enormous.

With the end of the war the wild-horse market vanished and the ranchers again became painfully aware that thousands of mustangs were roaming the range and eating grass the ranchers considered theirs, even if it was public land. A full-fledged war of extermination began, and thousands of horses were killed and ground into food for dogs, cats and chickens, or fertilizer. The hunt was so successful that by 1940 there were only a few thousand of the fleeter mustangs left in some of the most inaccessible and difficult terrain of the Rockies.

The horses were not entirely without friends, however. Wild Horse Annie—Mrs. Velma Johnston of Reno, Nevada—worked successfully for a Federal law forbidding the pursuit of wild horses in mechanized vehicles. She is now lobbying for a new law declaring wild horses an endangered species—a much more difficult job. State Senator George Jackson of Colorado introduced such a bill and saw it promptly defeated. Stockmen pack a lot of muscle in Colorado.

Such cowboys as are left and some oil-field workers still chase the wild horses for sport, and much of the pursuit is done by jeep, law or no law. The Bureau of Land Management has an even better idea for exterminating the horses. A program to be completed in 1971 will fence off what is left of open range into pastures of a few square miles. This will effectively keep the wild horses from water—which isn't the most pleasant way to die.

No discussion of endangered species can overlook the American bears—black, grizzly, polar or Alaskan brown bear. All face the likelihood of becoming a memory.

Victor Cahalane's 1965 survey for the New York Zoological Society estimated that there were only a few hundred grizzlies left in the states, most of them in two national parks—Yellowstone and Glacier. Alaska, he estimated, had about 17,000 to 18,000, probably 97 percent of the total.

The still unexplained rampage of grizzlies in Glacier Park, which resulted in the killing of two girls on the same night in two unrelated incidents, frightened some people into questioning the right of bears to remain in the parks. This attitude visualizes a wilderness area as something like a city park, manicured and patrolled.

Wilderness camping involves a certain amount of risk, of which encounters with grizzlies are probably the smallest. For the first 93 years in the history of national parks there were no fatalities involving bears, although 37,000,000 people visited 16 parks containing the animals. Other accidents were sometimes fatal. Drownings, falls, exposure—such mishaps caused 151 deaths in the parks in 1968 alone. As for bears, there is no question but that they can be dangerous. They look like comedians but they aren't. Each summer the park rangers nearly lose their reason when they see tourists trying to pet the bears, feeding them by hand, or putting the baby on the nice bear's back for a ride.

A short course in bear psychology appears to be a must for every visitor to a wild area. And the keystone of that psychology is that except under the most unusual of circumstances a bear is more anxious to avoid a man than the man is to avoid the bear.

Declining even more rapidly than grizzlies are the great white bears of the ice packs, the polar bears, which have been hunted since the whalers first sighted them more than 200 years ago. There are probably fewer than 10,000 of these bears left, with 1,200 to 1,400 killed annually.

Alaska was forced to cancel the spring 1970 hunting season because of the high rate of kills. Two new factors account for the sudden increase: oil and airplanes.

With the discovery of oil on Alaska's North Slope, many geophysical exploration teams have come in, with most of the men carrying firearms for "protection" against the bears. The North Slope is a treeless area on which an animal as large as a grizzly stands out like a mountain. In the short space of time that oil exploration has been going on in Alaska the grizzlies have been decimated.

While the use of airplanes to hunt bears in Alaska is permitted, an ex-fighter pilot named Joe Zentner had an even better idea. He mounted a semiautomatic rifle atop his Piper Cub and sighted it in at 150 yards. A Piper Cub is a slow-flying plane, and since the bear wouldn't be shooting back, the pilot could take his time, sail in slow and easy over the bear, and strafe it in real fighter-plane style. He killed 13 grizzlies this way, adding them to the 22 he'd killed earlier in the year by more primitive methods before the spoilsports in the Alaska Fish and Game Department stopped the fun.

To hunt polar bears the planes are used in tandem. Two planes fly out together, looking for bears on the ice. As soon as they find one the planes separate. One circles the bear; the other finds a spot to land. The circling plane then dives on the bear, forcing it to run in the direction of the landed plane. Driving the bear over the ice floes and ridges at top speed, it delivers the animal to the waiting hunter in a state of panic and exhaustion, unable to show fight—useless as that would be against a waiting high-powered rifle.

"Polar bears are stateless animals," wrote Alaskan Senator E. L. Bartlett in the *Audubon* Magazine, November–December 1966. "They live on the Arctic ice pack beyond the territorial limits of any nation, only occasionally and only in a few places ever coming ashore. Their welfare, their protection, are the responsibilities not just of America or Russia, but of all the Arctic nations."

An International Conference on the polar bear, held in Morges, Switzerland, discussed the new danger to the bears from airborne hunters. The Soviets expressed dismay that Americans, hunting by airplane, were frequently over international waters off the Siberian coast. The Russians for some years have banned polar-bear hunting completely, and have asked for a moratorium on hunting since 1965. Nevertheless the Alaska Fish and Game Department continued its approval of hunting by airplane.

In March 1970 Canada, which had permitted only Eskimos to hunt the white bear, suddenly lifted the curb. The Northwest Territories Council announced that it would permit Eskimos to sell their quotas to sport hunters. The hunter who pays an Eskimo $2,000 can now be taken by dog sled to hunt polar bears on the pack ice around Sachs Harbor and Resolute in the far north.

The council explained that the move was taken in the interest of the Eskimos. Previously an Eskimo might get only $200 for a polar-bear skin. This way he can get ten times as much. In other respects, said Dr. John S. Tener of the Canadian Wildlife Service, the quotas were reasonable. "We aren't pessimistic about polar bear numbers or stability of the population," he said. "We're still cautious." He conceded he didn't know how many polar bears there were.

The same ruling permitted Eskimos to hunt musk oxen again for the first time since 1917. There are about 7,000 of these shaggy animals left in northern Canada.

To the ecologist the Arctic is a special case anyway. Most people think of it as a rugged country—which it is—but the very harshness of the climate makes for a fragile ecology. The growing season is short, the frosts are deep and prolonged. Damaged natural resources take much longer to recover, if recovery takes place at all. Tracks left by vehicles in the tundra are still there years later.

Alaskan frontiersmen traditionally have acted like frontiersmen everywhere, as though nature were an adversary to be defeated. They thought the wildlife was inexhaustible, and their killing knew no reason or limit. The musk oxen were gone from Alaska by 1867, and most of the caribou by 1900. The valuable sea otters had been nearly exterminated by the Russians much earlier—which was one of the reasons they sold Alaska to the United States.

Alaska offered bounties for eagles, seals (they catch salmon), coyotes, wolves and wolverines. By 1920 the seals of the Pribilof Islands were nearly extinct, and in the following years the decline of the walrus, salmon, bear and wolf has destroyed the myth of Alaska's limitless abundance.

Man is oddly ambivalent about animals anyway. Most people "love" animals. The zoo is a favorite visiting area, animal movies do well at the box office, pet shops are big business, and almost everyone stops to pat the policeman's horse, feeling a vestigial kinship with the wild as he does so. Yet the slaughter goes on. Hunting magazines are filled with first-person stories by hunters who rhapsodize about their love for the magnificent wild goose as it wings down out of the sky—and as they blast it with a load of shot.

In the Gulf of St. Lawrence, as the spring ice begins to soften, harp seal pups have been clubbed to death for years. By ship, plane and helicopter the sealers, called "swilers" in Newfoundland dialect, swarm over the ice to kill baby seals.

The harp seal pups cannot swim at birth, so the mother seals come out on the ice to give birth to a snow-white baby with great dark eyes, which they nurse on the richest milk in the world, about 50 percent butterfat. On this diet the babies grow from about 20 pounds at birth to 90 or 100 pounds in two weeks.

The white fur of the harp seal pup has a high market value, so to avoid damaging the skins, the swilers club the young over the head and skin them on the spot, dead or not, while the panic-stricken mothers escape into the water or helplessly watch. In an orgy of killing, the men are smeared with blood and the ice is stained with great red blotches as they hunt down one group of seals after another.

About 50,000 seal pups, the legal limit, are slaughtered in three to ten days, after which the hunters move on to the North Atlantic ice pack, called the "front," where there is no limit but time. Here the Canadians are joined by Norwegians for a 13-day season in which they kill about 200,000 seals.

Bad publicity about the cruelty of the hunts began as early as 1957, coming from Canadian humane societies. By 1962 newspaper stories began to appear, replete with gory details. The clincher came in 1964 with two films made by television crews of the Canadian Broadcasting Corporation. A sequence in one film showed seal pups being skinned alive; the Canadian Department of Fisheries said it had been staged. Perhaps it was, although it is a little difficult to imagine the swilers being persuaded to stage an event so obviously self-incriminating.

Public reaction to stills and motion pictures showing the vicious clubbing and skinning of these beautiful and appealing pups was violent. Angry letters and petitions poured in on the Canadian officials. "If they only looked like alligators we'd have no trouble," mourned one official.

Even more persuasive was a spontaneous boycott of seal furs that dropped the price and the income of the seal hunters. It spread to others besides the harp seal hunters. *Time* Magazine reported that the income of Canadian Eskimos depending on the seal hunts fell from $750,000 to $95,000 in five years.

Alarmed by this damaging backlash, a furriers' group began running ads in newspapers and magazines deploring the misguided efforts of humane societies in condemning the annual seal harvests. The ads were unfortunately somewhat obscure, because a clear distinction was not made between the harp seal killings and those of the Alaskan, or Pribilof, seal, which was the real object of discussion. To have made the distinction would have been to condemn, at least by inference, their confreres in the fur business. So the ads befogged the issue a little more by a defense of something not really under attack.

Apparently the Pribilof seal has come back from the thin edge of extinction, for the Audubon Society says:

> Humane organizations—who have allied themselves with conservationists on many important causes—may wish to argue against the seal harvest in Alaska on moral grounds, and they have a clear right to do so. It should

be understood, however, that the Alaska fur seal, numbering 1,500,000 animals on the Pribilofs, is by no stretch of the imagination rare or endangered. Indeed, the saving of the fur seal from extinction—and its longtime management—is one of the great conservation stories of our history.

From the humane standpoint the thought of another 120,000 intelligent and beautiful animals being clubbed or shot to death is far from appealing, even with the assurance of the Fish and Wildlife Service that these are "mostly surplus bachelor males." Even if the sale of skins brings in $28,000,000 a year to the Federal and Alaskan treasuries—always a potent argument.

The principal furore, however, has revolved around the harp seals and the rarer hood seals of Greenland. The hood seals have declined by half in the decade of 1950 to 1960—from about 3,000,000 to 1,250,000. The birth rate has dropped from 750,000 to about 315,000. Mortality of the pups has increased from 30 percent to 75 percent. Graphically showing the decline, more than 500,000 pups were "harvested" in the first half of the nineteenth century, nearly double the number now being *born.*

As to the harp seal, the Canadian Fisheries Research Board believes it is stabilized at the present rate of kill. But Dr. Douglas H. Pimlott of the Canadian Audubon Society says, "The industry is greedy, it is overcapitalized and it holds a very short-term view of the resources. If it prevails in its demand for higher annual quotas than the seal population can sustain, the harp seals of the Front herd are likely to be exterminated within this century."

The seal slaughter is commercially motivated if nothing else, and wrings a "this hurts me more than it does you" attitude from the fur industry, however hypocritical that might be. But the slaughter of the California sea lions has no motive whatever except man's undiminished love to kill.

The Santa Barbara Channel, where oil leaks have already taken a toll of birds, fish, porpoises, seals and sea lions, is now yielding up more dead seals, goats, sheep and birds, riddled by bullets.

The channel contains eight steep-sided rocky islands that are breeding places for wildlife. It has become a temptation for gun-happy delinquents to go out in boats and shoot at the birds and animals. There are no laws protecting seals, sea lions or porpoises in California, and shooting them is encouraged by commercial fishermen, who resent the animals' eating fish that, of course, belong to the fishermen. The California Fish and Game Department supports the fishermen, not the wildlife.

No one knows how many sea lions have been shot, because many of them are killed in the water, where they sink out of sight. What is obvious is that their numbers are sharply declining. Volunteers in a group called the Fund for Animals, headed by Cleveland Amory, have organized boat patrols to try to protect the sea lions. Their only weapon is persuasion, but the "sportsmen," far from being embarrassed by the aimless killing, tend to be pugnacious and self-righteous, so the mission is not without excitement.

Sea otters, an entirely different animal, are protected by law but are being shot illegally by abalone fishermen on the California coast. There are only 700 or 800 sea otters left in California, though about double that many are in the Aleutian Islands National Wildlife Refuge. The sea otter possesses a fur that is unbelievably dense and soft. The animal depends on it entirely for warmth as, unlike the seal, it does not build up a layer of insulating fat.

At their low point the sea otters were down to about 500 individuals before protective measures went into effect. Now the abalone fishermen are pressing the state Fish and Game Department to trap the California colony and move it elsewhere, claiming the otters are eating the abalone.

It appears that the abalone catch has dropped from 1,400,000 pounds in 1957 to 414,000 pounds in 1969. Can 800 otters eat a million pounds of abalone a year? Dr. Melvin Odemar, a fish and game biologist, believes they can (*New York Times*, May 10, 1970). He says a sea otter can eat 45 to 75 pounds of shellfish in three or four days. He calculated that one small herd of 97 otters must have eaten between 627,800 and 1,150,000 pounds of abalone in a year. Of course otters eat other shellfish than abalone, which probably accounts for the unusual spread in this estimate, almost double the minimum.

Speaking for the otter is Mrs. Margaret Owings of Big Sur, founder of Friends of the Sea Otter. She says, "The truth is that the industry has depleted this resource by overfishing. Why do they have to blame the otters?"

Of course this is a very old story. Wolves were blamed for the decline of the caribou in Alaska, cougars blamed for killing the deer, crows blamed for killing the ducks. The truth is that no natural predator can exterminate a species; that job is done by man, for no predator commercializes his catch as man does.

The sea otter is one of few animals known to use tools like a human. An otter will dive with a stone, which it uses like a hammer to knock loose the grip of an abalone's suction foot on a rock. Sea otters have been observed floating on their backs in the water with a rock balanced on their chests and a clam or other shellfish held in their paws, which they raised in the air and brought down on the rock, hammering away until it cracked.

The list of vanishing animals is a depressingly long one—the black-footed ferret, the Sonoran pronghorn antelope, the Eskimo curlew, the osprey and eagle, the black-crowned night heron, the cahow, the lynx, the prairie dog, the coyote, the Florida Key deer, the fabulous quetzal—a bird sacred to the Mayans and Aztecs, now the national bird of Guatemala but still vanishing from Central America. Some are the victims of hunters, some of pesticides, some merely of urbanization.

"Nothing lives here," wrote John Barbour of the Associated Press in the Miami *Herald*, May 26, 1968, "except man gives it permission to live. Nothing survives unless it serves him to have it survive."

Do we have an obligation to other forms of life? Perhaps it isn't enough to save an animal or a species because it is picturesque or valuable or of interest to science. If man is ever to achieve more than the lip-service reverence for life he spouts so glibly, it must be in the direction of recognizing the right of other creatures to life.

This is not an abstraction, not merely sentimental or idealistic. It is deadly practical. For if we continue to ignore it, to the list of endangered species we must surely add one more—man.

PART IV

The Blight of Civilization

CHAPTER

7

The Ecology of the City

OUR society of abundance has achieved a new height in disposability: we are now throwing away whole cities.

In New York, Philadelphia, Detroit, St. Louis—many of the older cities—landlords are simply walking away from their houses. Whole blocks are being abandoned; entire neighborhoods are ghostly no-man's-lands of ruins in the making.

"Parts of these cities are so empty," a Federal official is quoted as saying, "they look as if someone had dropped nerve gas."

In the human attempt to find someone to blame there has been soul-searching and name-calling. The problems of the city are so many—strangulation by political bureaucracy, the flight of the middle class to the suburbs, spiraling taxes, failing services, pollution, garbage, congestion, noise, dirt, crime, hostility, insensitivity, and the disappearance of the common amenities—the list is frightening.

Shortly after his inauguration President Nixon told a meeting of mayors, "I want you to understand that problems of the city will be on the front burner of this Administration."

Apparently someone forgot to light the gas, for in two years the model-cities program had been scuttled and Federal aid had all but dried up. Urban renewal and employment had been cut. It came as a shock only to the Administration to discover in the spring of 1970 that unemployment among young blacks was 30 percent nationwide, and as high as 45 percent in Los Angeles.

The hopelessness and helplessness of some young progressive mayors—Cavanagh of Detroit, Allen of Atlanta and Naftalin of Minneapolis—must have been overpowering to force them to quit as they did. An acidic, anonymous jingle ran:

> Why are the mayors all quitting?
> Why are the cities all broke?
> Why are the people all angry?
> Why are we dying of smoke?
> Why are the streets unprotected?
> Why are the schools in distress?
> Why is the trash uncollected?
> How did we make such a mess?

Rallied by Lindsay of New York, mayors have tried to bring political pressure on the Federal government and their state governments to provide more financial aid. "At this rate," said the mayor of one city bitterly, "we're all going to be like the people in the Nazi concentration camps, scrambling against each other for the Federal crumbs necessary even to stay alive" (*New York Times,* May 22, 1970).

The scramble to stay alive has already resulted in a pathological insensitivity, glaringly epitomized in the case of a girl who was stabbed to death on a city street while 60 witnesses who heard her screams failed even to phone the police because they didn't feel personally involved.

The hope that Federal money will solve the problems of the city is as much an illusion as that our technology is equal to all the problems it has created.

"Massive increases in Federal aid," says Edward N. Costikyan ("Cities Can Work," *Saturday Review,* April 4, 1970), former leader of the Democratic Committee of New York County, "would not solve a city's problems, but rather would be quickly ingested by the money-consuming monster that city government has become."

He points out that in New York the cost of providing essential services goes up every year by about 15 percent, while revenues, despite the desperate measures employed, rise by less than 5 percent. Even liberal infusions of Federal money will simply be absorbed by the proliferating bureaucracy.

Although politics, crime, racial conflict, social problems and all the rest can fairly be said to be part of the city's ecology, they are beyond our scope here. We will have to stay with a fairly superficial look at technology, pollution, urban sprawl and planning for space.

What is totally lacking in all the glittering projections of technological miracles promised for the future is a balanced sense of values. We know that the administration has earmarked $4,000,000,000 to clean up the water over the next several years and that this is little more than a gesture, since it is more likely to require $20,000,000,000 a year for the next several years. So in a rough way we know what open space, clean air and clean water will cost, but we cannot calculate its cash value in terms of human life. The shapers of our society unfortunately have too often been the men looking for a fast buck, and they have left their mark.

Phillip Berry, president of the Sierra Club, tells of a White House meeting with President Nixon in which the President invited support from the conservation societies for his environmental program. He thought the "current ecology fad" wouldn't last, that students would soon be onto something else, and the public was too fickle to ensure support in getting the administration's program through Congress.

> I drew the first Presidential frown [said Berry] when I suggested our goals go considerably beyond the President's and the distance between us couldn't be measured in dollars alone. . . .
>
> I explained that conservationists believe limitless expansion of the economy is an outmoded idea which is causing us to eat into our basic and irreplaceable capital—the environment itself. I noted conservationists believe we can't any longer measure progress by growth of the gross national product alone, or even primarily in such terms. On the contrary, conservationists are urging we measure progress by addition to the quality of life. I argued that zero population growth should be a national priority, in keeping with the urgency of the population problem. Finally I said conservationists want fundamental changes in the ways we think about the environment; they want those who would change it to carry the burden of proof that the change will not be adverse; they want effective controls over now rampant technology; they want adoption of the land ethic and the changes in life styles it implies. I said I hoped the President could adopt those ideas.*

The ecology fad that so unexpectedly swept the country was con-

* *Sierra Club Bulletin,* April 1970.

cerned primarily with pollution, and with good reason. Pollution is the visible symptom of technology gone awry, or not gone far enough. Some 200,000,000 tons of contaminants are pumped into the air each year. Every *second,* 2,000,000 gallons of sewage are poured into our rivers and bays. The dust fall on Manhattan Island is 80 tons per square mile per month.

Small wonder biologist Barry Commoner greeted the students at the University of Michigan teach-in by saying, "Congratulations. You are the first generation to carry strontium-90 in your bones, DDT in your fat, iodine-131 in your thyroid glands and coal and asbestos dust in your lungs."

The greatest single source of air pollution is the automobile—60 percent of the total as against 17 percent for industry and 14 percent for electric power plants.

The exhaust gases from cars contain a choice collection of irritants: carbon monoxide, sulphur oxide, hydrocarbons, nitrogen oxides and lead. Sunlight has a photochemical action on this mix, the hydrocarbons combining with the nitrogen oxides to form ozone and other poisons in a witches' brew we know as smog.

Milestones in history are easy to spot after you are past them. You can go back as far as 1930 and the first air-pollution disaster in the Meuse Valley in Belgium. But take 1948, Donora, Pennsylvania.

On that 27th day in October the temperature dropped and a high-pressure system over Donora had created an inversion—warm air above a layer of cold air at ground level, so that the pollutants could not rise and be dispersed.

A steel mill, a zinc mill, auto exhausts, trains, boats and chimneys poured their usual load of exhaust into the air. It began on Wednesday. On Thursday people with a history of respiratory trouble began to be ill. All night the ambulances rode, and by Friday the hospitals had suddenly, frighteningly, filled. By Saturday it was all but over as rain and wind tore the smog to shreds and blew it away. But 20 people were dead and 5,910 were ill—42.7 percent of the population of Donora.

Blame the mills or the car exhausts? "In those days," said a Donora man, "it was pretty lonely to be against air pollution."

Thanksgiving Day, 1966. It was a mild day in New York, and "inversion" was still a new word. The sulphurous umbrella that hung over the city had lowered, and a yellow curtain seemed to have closed off the long vistas of the streets. People began to cough. And by the time the inversion had lifted, 168 were dead, most of them elderly, with a known respiratory problem.

These were not large catastrophes measured by London's four days in 1952 when 4,000 over the normal death rate died and the reported illness among respiratory and cardiac patients doubled. But they were grim warnings.

There are people who believe that the narrow canyons of our city streets are not designed for automobiles, that a sensible solution to smog and to concentrations of carbon monoxide five or ten times higher than the danger point is to ban cars from the congested downtown areas. Others are confident that a pollution-free engine can and will be produced.

Regulations now call for the reduction of hydrocarbon and carbon monoxide emissions during 1971, and nitrogen oxides by 1973. But an item in the Washington, D.C., *Evening Star* is worth noting: "The nation's big auto makers are very carefully hedging their promises on how early they can install sophisticated devices to meet proposed Federal air pollution standards."

At a House Public Health and Welfare subcommittee meeting in April 1970 both Ford and General Motors said "there were no promises." The auto makers passed the buck to the oil companies, blaming lead in gasoline for the problem. Other testimony before the committee indicated that even the anti-pollution devices already installed on cars generally became ineffective after driving 10,000 miles.

What is expected to happen, says the National Air Pollution Control Administration, is that pollution from cars will diminish to some extent in the seventies but will be increasing again by 1980 simply because there will be more cars than ever. Manufacturers, with billions invested in present technology, have shown anything but enthusiasm for developing replacements for the internal combustion engine.

Nor does pollution remain within the city even if it is produced there. Wind currents carry it to regions once relatively clear.

Millions of ponderosa pine trees in the San Bernardino Mountains 60 miles from Los Angeles are being killed. La Jolla, 80 miles south of Los Angeles, is being invaded by a dark cloud of pollution drifting from the north. In Hendersonville, North Carolina, the famous pines are dying as fumes from the paper mills pervade the air. Asheville, once the fashionable center of the Great Smoky Mountain region, is black with grime and its air acid from the pickling liquors of the paper plants. Florida's orange trees are being killed by sulphur oxides from fertilizer plants; pines in the Spokane area have been damaged by fluoride pollution from an aluminum plant; and thousands of acres of white pine in the Cumberland Plateau of Tennessee have been killed by stack emissions from the TVA steam plant and the Oak Ridge laboratories of the Atomic Energy Commission.

On Whiteface Mountain in New York, one of the cleanest areas in the nation, pollution is drifting up through the valleys. In May 1970 the Atmospheric Sciences Research Center on Whiteface reported a pollution level of 5,000 to 7,000 particles per cubic centimeter of air, a level approaching that of a major city. The clean air level of Whiteface had been habitually under 1,000 particles; 1,000 to 5,000 is normal, while industrial city air may contain as many as 50,000 particles.

There is reason to believe that government standards for clean air in the cities are too low. On a day in New York when the city rated the air as "acceptable" the Citizens for Clean Air reported: "Sulphur dioxide comes chiefly from oil burners and electric power stations. According to the Federal Government, yearly average levels greater than 0.04 parts per million are associated with chronic bronchitis and many lung diseases. Today the highest 24-hour average was 0.07 parts per million in Sheepshead Bay."

Less well publicized, but no small factor in the ecology of cities, is ear pollution. The level of noise has doubled since 1954 and now in New York regularly exceeds 85 decibels, a point at which loss of hearing can be expected if it is maintained.

Nor is hearing loss the only penalty of decibels. Noise causes physiologic changes generally associated with stress. It causes blood vessels to constrict and blood pressure to rise, diminishes stomach secretions and causes an increased flow of adrenalin, with associated cardiovascular and glandular changes.

Steelworkers in noisy areas become irritable and argumentative not only on the job but at home, and workers in noisy jobs generally suffer more from fatigue and have more neuroses than the average.

When sound reaches a level of 165 decibels it will kill small animals like mice, rabbits or cats. The energy of sound is converted into heat in their bodies. It is estimated that 175 decibels will kill a man. So far the highest noise recorded is 155 decibels, produced by a jet plane with afterburner taking off from an aircraft carrier deck. But it might be instructive to know that a subway train produces about 95 decibels, a motorcycle about 110 decibels, and a commercial jet on takeoff about 150 decibels.

Which brings us to an interesting point. The Federal Aviation Authority considers 100 decibels the point at which most people will find the noise intolerable. The new supersonic transport, the SST, will generate sideline noise at 1,500 feet of nearly 130 decibels. Faced with this unpalatable fact, the FAA simply raised its level to 108 decibels, and continues to support the development of the new plane.

More serious, the SST produces a sonic boom in flight, a shock wave caused as the plane exceeds the speed of sound. The shock wave is the equivalent of a heavy explosion that breaks windows and cracks walls. It takes the form of a cone reaching the ground from the nose of the plane and spreading about 25 miles to each side, so that the boom is a continuous one traveling with the airplane over an area about 50 miles wide.

The boom has already shattered the nerves of populations in Oklahoma, killed 2,000 mink in Minnesota, driven a herd of cattle off a cliff in Switzerland, and killed three men in France when their farmhouse collapsed upon them. Bailey Smith of Oklahoma City won a verdict of $10,000 from the U.S. Court of Appeals when a sonic boom split his house in two.

The Air Force has already paid out more than $1,500,000 in damages caused by the sonic booms of military aircraft, and another $2,500,000 in claims were filed just in 1968. Damage has been done even in the wilderness. A whole section of cliff in the Canyon De Chelly national monument in Arizona collapsed and fell on a section of irreplaceable prehistoric Indian cliff dwellings.

The undeterred head of the supersonic transport program, William M. Magruder, offered his opinion that the public has been confused by numerous half-truths about the SST and that the finished plane would be quieter than conventional jet planes. He conceded that the noise would be "three to four times as annoying" as that of jet planes, but it would be largely confined to the airport.

And in any case, he believed that technological skills would reduce the noise level by the time the SST was ready in 1978.

One argument for the giant plane is that the government has already sunk $700,000,000 into it and would lose all that money if the project were abandoned now. Of course if the government spends several billions more and the plane then turns out to be a white elephant after all, the loss presumably would be easier to bear. Another argument heard is that if the project were canceled it might well have serious repercussions for the Boeing Company and the city of Seattle. To which Congressman Reuss of Wisconsin replied, "Why in the name of common sense do we not put Boeing to work making a mass transit vehicle and General Electric to work producing a pollution-free engine for it?"

Allied to pollution is the problem of waste disposal. It can be simply stated: We are running out of places to dump garbage.

Every day every American is responsible directly or indirectly for seven pounds of trash—about 530,000 tons altogether. Only 12 percent of it is garbage; a lot of it, unfortunately, is indestructible—materials like glass, ceramics, concrete, brick, metal, and plastics. Garbage can be burned in incinerators, adding to the air pollution, but the indestructible materials, and a lot of paper that could be recycled, are dumped.

The former sanitation commissioner of New York, Samuel J. Kearing, Jr., describes in *New York* Magazine for April 13, 1970 his introduction to the job in 1966. He learned that the Sanitation Department was dumping its garbage on Staten Island, the last open rural area in the city. "And when Staten Island is filled, where will we dump the garbage next?" he asked.

"We'll fill Jamaica Bay."

"But that's a park and a wildlife refuge."

"It doesn't matter, Commissioner. If you can't burn it and you can't export it, then Jamaica Bay is the only place left."

One can find a modicum of sympathy for city officials caught in an inexorable squeeze between the absolute necessity of getting rid of the garbage and the various uproars it engenders. Jamaica Bay is currently a wildlife refuge only because Staten Island, a Republican stronghold, lost the unpopularity contest with a Democratic administration. The result was that 1,500 acres of salt marsh at Great

Kills and 3,000 acres at Fresh Kills are now buried under millions and millions of tons of garbage.

Paradoxically enough, the man who turned back the garbage tide at Jamaica Bay was Parks Commissioner Robert Moses, a most unlikely champion, whose interests have always run to manicured parkways rather than wild areas and who has never been heard to express any fondness for bird watchers.

However, Jamaica Bay is hardly inviolate. There is dumping at Spring Creek and Edgemere on the borders of the bay; filling by developers has hampered the flow of fresh water into the bay; and an airplane runway from Kennedy Airport has blocked circulation in the eastern end. If Floyd Bennett Field in Brooklyn is converted to a cargo center as contemplated, channels for oceangoing ships would then be dredged into the bay.

There is also a proposal on the drawing boards for a new Gateway National Recreation Area that would take in Jamaica Bay. Since the area is marsh, the nature improvers have suggested filling in certain boggy islands in the bay to elevate them into "useful" land, implying that land that is merely the nesting ground of herons and egrets is hardly useful. The fill? Garbage, what else?

Looming somewhere in the future is the long-term threat. What, indeed, happens when Staten Island is filled up?

In one way, according to Mr. Kearing, New York is in a better position than most localities. The dump at Fresh Kills is rated as a "sanitary land fill" because the refuse is covered with six inches of earth. Of 12,000 other landfill sites in the United States surveyed by the Public Health Service, 94 percent were rated as "unacceptable and represent disease potential, threat of pollution and blight."

About a third of New York's garbage is incinerated, the rest dumped. It had been dumped at sea; in 1933 this was stopped by an order of the Supreme Court. Now only sewage sludge and chemical wastes are still carried by barge out into the ocean and dumped off Ambrose Light.

More than one housing development has been built on a garbage dump, but a new wrinkle is Mount Trashmore in Du Page County, Illinois. The county had run out of dumping ground, but there was a marsh, somewhat the worse for wear, known as the Badlands. A waste-management specialist from the University of Chicago suggested killing three birds with one plan: fill the swamp with garbage, making a mountain out of it, and convert it to a ski run.

The project was started in 1965 and is to reach a height of 125 feet when completed, the highest elevation in the county. It will have five ski slopes and six toboggan runs. The hill will be built up in honeycomb fashion, each cell a square yard of garbage surrounded by a foot of packed clay to seal it against seepage into the ground water.

From Berlin comes reports of a "Mount Junk," a 360-foot hill built of rubble from wartime bombings, used as a ski jump, toboggan run and general lookout over the city. Its success has apparently stimulated New York City Council President Sanford D. Garelik to outdo it. He proposed what *The New York Times* called a "Grand Teton of garbage," a 2,500-foot mountain of refuse to be erected in the Bronx as a ski resort "second to none."

Mount Garelik ran into the usual derisive opposition, but hard on its heels Chicago announced plans for a 1,000-foot garbage mountain to be spread over five square miles which would take more than 25 years to complete. It may be the wave of the future, for there seems to be no practical limit to the number of such mountains that can be built.

A textbook study in the no-solution dilemma is the struggle over San Francisco Bay.

To coin an advertising slogan, San Francisco has been cited by more independent surveys as the most charming city in the United States. Yet San Francisco's historian Harold Gilliam has said, "The truth should be faced, heretical though it may be; man-made San Francisco is an ugly city. With the exception of a very few well-designed buildings, most of the works of man here are unworthy of the site. It is nature that gives the city its distinction—the hills, the climate, the surrounding waters. Without them San Francisco would be almost indistinguishable from a hundred other cities."

The bay, said former Interior Secretary Udall, is "the greatest single resource in this region." Aside from the fact that it is the great scenic feature of the Golden Gate, the bay has practical virtues as well. It is an effective air-conditioner for San Francisco, Oakland, Berkeley and the other bayside communities, providing cool summers, mild winters, and pleasant breezes. It supplies oxygen and reduces smog. Like all tidal waters, it is a prolific breeding ground for fish, waterfowl, animals and shellfish, some of which have high commercial value.

It would seem that residents of this fortunate area would prize such an asset beyond compare. But what they really wanted was to fill in the bay.

During the fifties and sixties, one after another of the bayside cities developed plans for expansion—outward into the bay. They were already dumping their garbage into it—1,600 tons a day from San Francisco alone. They had already created some fills; Treasure Island in the middle of the bay, the San Francisco and Oakland airports—all made on filled land. And all around the edges, marinas and housing developments were chewing away the shoreline. At one time there were 34 different fill projects, estimated at 16,000 acres. Oakland proposed to fill in the bay right out to Treasure Island in the middle.

The Army Corps of Engineers reported that two-thirds of the bay was 12 feet deep or less, and thus "susceptible of reclamation." Translation: Let's fill in the whole thing.

Berkeley produced a plan that showed it could double its size by filling in 2,000 acres of the bay. Foster City planned to create 2,600 acres by dredging 18,000,000 yards of sand from the bottom. Sausalito pondered a huge fill to accommodate 20 buildings with apartments, hotel rooms, shops and parking lots. When the city council put a moratorium on fills, the developers brought suit, saying they had bought underwater lots and had a right to fill them.

By 1960 a third of the bay was gone, shrunk from 700 square miles to 400. The once clear blue waters were stained and polluted. The shoreline marshes were junkyards and had dwindled from 300 square miles to 75. Fills and dikes had raised barriers against the normal flushing action of the tides; the oxygen content was down, sewage pollution up, and fish dying. Ninety percent of the bay was closed for shellfish gathering and 40 percent closed for water sports.

It was the Berkeley plan that galvanized a scattered opposition into closing ranks and making a fight. Mrs. Clark Kerr, wife of the president of the University of California, enlisted two friends, Mrs. Donald McLaughlin, wife of the chairman of the University Regents, and Mrs. Charles Gulick, wife of a professor. They went for help to the conservation clubs and got backing from the Sierra Club, the Audubon Society and the Save-the-Redwoods League.

Out of this coalition came a Save San Francisco Bay Association, which won backing from the late Senator J. Eugene McAteer. With Nicholas Petris of Oakland, then an assemblyman, McAteer drafted a bill calling for a three-year moratorium on landfill and establishing a Bay Conservation and Development Commission to make a detailed study of all the factors affecting the bay.

Some of the things they found were illuminating. One enterprising soul had bought underwater land for one dollar an acre, eventually selling it to a developer for $3,000 an acre. The developer filled it to five feet above high tide at a cost of another $3,000 an acre and then sold it to a builder for $45,000 an acre.

In the five years from 1957 to 1962 San Mateo County had bulldozed 10,000,000 cubic yards of earth into its waterfront, and planned to push in an additional 636,000,000 cubic yards to fill 23 square miles of water.

The McAteer study, completed in 1969, declared that "any filling is harmful to the bay" because it upsets the ecological balance, increases pollution, and reduces oxygenation.

The developers were outraged. One entrepreneur told a legislative committee that he naturally expected a law to have some teeth, "but this one has fangs!" Another builder accused the conservationists of being more concerned with birds than with poor Negro kids. A little homework on the part of the committee revealed that the housing project in question was planned for homes in the $20,000 to $60,000 range—which would have excluded quite a few poor Negro kids.

The Save San Francisco Bay Association worked for three years, expanding its membership from 5,000 to 22,000 and getting 200,000 signatures on a petition to present to Governor Reagan. When the governor declined to receive the petition in person, the by-now publicity-wise group strung all the papers together in a three-and-a-half-mile-long strip and tastefully decorated the capitol steps and lawns with it. (They picked it all up afterward.) They sent a bag of sand to a state legislator with a label that read, "You'll wonder where the water went if you fill the Bay with sediment." Undignified but effective.

The bill for a moratorium passed, with a provision that the commission have three years to purchase the privately owned shore lands that it found necessary to protect the character of the bay. The commission has no funds for such purchases, however, and Governor Reagan has made it known that he does not believe the state should pay. In the end, if anything is done it may be in the form of a special bay income tax.

What is left of San Francisco Bay is saved for the time being. It is something of an ecological victory for the citizens of the bay area (the disappointment of developers notwithstanding), because filling in the bay would surely have changed the climate and removed the chief attraction of the entire area.

Like all conservation victories, it is a temporary one. A project backed by the Chase Manhattan Bank contemplates shearing off the tops of the San Bruno Mountains south of San Francisco, creating a plateau for building, and then using the removed earth and rock to fill in the bay at the foot of the mountain. A hundred and fifty feet of the mountain would be removed, and the fill would be the size of Manhattan Island, with shopping centers, skyscrapers, industrial parks and 80,000 homes.

"The battle is not now, and never will be won," said Joseph Bodovitz, executive director of the commission. "Maintaining a high degree of continuing public interest is crucial."

Where a natural barrier, such as a shoreline, does not intercede, cities tend to spread, most of the time unfortunately, in an ugly, uncontrolled sprawl. The spread is due less to pressure from the dense inner core than to the efforts of developers, whose business it is to create new areas and urge people to move into them. And, regretfully, this is the atmosphere that has shaped our values. The real-estate speculator and the road builder have designed our cities. Buildings are created not with the goal of adding beauty or dignity to the cityscape but to provide the most profit to entrepreneurs who hardly gave a thought to those who might live there. The result has been the dehumanization of our cities—and the flight to the suburbs by those who could go.

Density creates problems of many kinds. The Greater New York metropolitan region covers an area of some 6,900 square miles in 22 counties of New York, New Jersey and Connecticut, with a total population (1960 census) of 16,137,000 residents. The density works out to 77,000 people per square mile in Manhattan and tapers to 200 people per square mile at the far edges of the region. The average for New York City is 25,000 per square mile, and the average for the entire region is 2,337 people per square mile.

The congestion can become intolerable. A famous experiment performed by Dr. John B. Calhoun of the National Institute of Mental Health with laboratory rats showed emotional collapse under severe crowding. Normally sociable, the rats began to fight among themselves, destroyed their young, withdrew from normal sex patterns entirely or became homosexual, and showed other symptoms of mental illness. Among humans the rate of mental illness is about one in ten, but in the cities the rate is closer to one in four.

Still, not everyone objects to high density, at least where the living conditions are tolerable. The large city offers excitement and vitality, contact with top achievers, a concentration of talent, a huge variety of experience and a great choice of activity. The city makes certain services more efficient, supports better shops, greater cultural facilities and mass transportation. Educational facilities are better, the incidence of disease is lower and medical services generally better, the material standard of living for all but the very poor is usually higher.

The trend toward urbanization therefore continues as people leave rural areas for the larger communities. The interesting statistic here is that while 70 percent of the American population now lives in urban and suburban communities, they occupy only about 1 percent of the land. Even if the population exceeds 300,000,000 by the year 2000, the estimates are that it will occupy only 2 percent of the land.

How, then, have we managed to spoil, scar, and sterilize so much of our landscape? Even where it is not occupied we have used the available space badly, even viciously. The supply of land has until now seemed inexhaustible to a people clinging stubbornly to a vanished frontier. The pattern was always to use up and move on—until there was no more place to move to. Yet communities continue to sprawl without plan, and the helter-skelter collections of shacks, roadside blights and industrial eyesores have metastasized like malignant colonies reproducing themselves endlessly.

The ugliness and deficiencies of our urban areas come from the failure to plan for open space that would have made the high densities tolerable. Even more inexcusable, the failures of the cities are being repeated in the suburbs, which are spreading in the same haphazard fashion, without provision for parks, play areas or recreational facilities, particularly for young people. The result has been called a "slurb"—an endless slum city where the traveler can go on indefinitely through street after street with no change of scene and even without the realization that he has passed from one town into the next.

Relief from the immense, encompassing anthill can be achieved by the use of open space to provide freshness of scene, recognizable landmarks, a feeling of coherence, a sense of individual communities. Such spaces can take many forms—a plaza, a terrace, a park or garden—what Kevin Lynch of the Massachusetts Institute of Technology calls "an urban analogue of wilderness."

Some opening awareness of this need, however feeble, is beginning to make itself felt. In New York the city administration has built, here and there, tiny oases called vest-pocket parks. Some of the new buildings are being set back—instead of crowding the sidewalks for the last inch of space—to create open malls, pedestrian walkways, even some garden plantings.

It doesn't matter if these are slightly incongruous copies of old and mellow European artifacts, an imitation of an Italian galleria, a Parisian sidewalk café, an arched Spanish courtyard or a Viennese cobbled square, even a replica of an English mews. The space, the imprint of character, is what counts. Where the blare of traffic is banned and pedestrians can stroll as in Copenhagen's Strøget or the new pedestrian mall in the heart of Stockholm's business section—even Lincoln Road in Miami Beach—life is added to the city. And growing out of this thinking, this use of small open spaces, is the idea that we should not fight density by diluting and decentralizing but by creating *more* density. Controlled density.

This concept makes a good deal of hard, if unpleasant, sense. It seems inevitable that we are going to have a larger population, wherever the eventual leveling-off takes place, but we certainly are not going to have more land. Unchecked sprawl simply has to give way to more efficient use of the available space. In the crowded cities this means higher densities, with the ingenious use of small spaces to simulate larger spaces.

This planning has begun, with several American cities tearing down and rebuilding their central business sections, designing small open spaces to separate or surround the buildings so there is a visual effect of space. "Man cannot live by elevator shafts alone," said Frederick J. Close of the Aluminum Company of America, a company involved in housing. "He has to have an occasional view of a sunset, a cool morning breeze and a mud puddle" (*Fortune* Magazine, February 1970).

So the Yerba Buena project in San Francisco will have a convention center, an exhibit hall, a sports arena, a theater, parking garages, a downtown air terminal, an Italian culture center, office buildings and a hotel.

Seattle's downtown section is being redone to coordinate mass transit with cars, pedestrian malls, stores and plazas, all designed on multilevels with open spaces and plantings.

Crown Center, a project in Kansas City, is being built on 100 acres entirely by Hallmark, the greeting-card company. It will contain office space, apartments, specialty shops, movie houses and restaurants. Philadelphia's center has been redone and its historical sites restored. Boston has replaced the old Scollay Square with a new government center of city, state and Federal buildings. Baltimore has rebuilt its downtown section around a pedestrian plaza where symphony concerts are performed on summer nights.

All this helps, at least to keep the city a viable commercial force. But rebuilding the business section does little for the ghettos, where abandonment is common, and the outer belts of the cities, where sprawl continues.

The American ambition of owning a house and a private plot of land, however small, has resulted in too many machine-made developments, of which the ultimate so far has been Levittown on Long Island. The sheer size of this development must be seen from the air to be fully grasped and to appreciate Kingsley Davis' comment on large communities that they suggest the behavior of communal insects rather than of mammals.

The pity of it is not merely the unrelieved monotony of thousands of identical houses and identical plots but that this is a prodigiously bad use of land. In more expensive developments, zoning for larger plots has resulted in fewer houses to the acre, but this merely used more land for fewer people.

Urban renewal has had to cope with the problem of relocating slum tenants and the high price of city slum areas, valuable even if filthy and decaying. Professor Nathan Glazer of the University of

California at Berkeley estimates that urban renewal in New York City has cost the taxpayer $1,000,000 an acre—the difference between the price paid to the owners of the land and the price the developers paid for it.

The size of the problem has led some observers to feel that the older cities might as well be written off and entirely new cities planned from scratch. The rash of blue-sky projects has often reached into the science-fiction realm: the linear city stretching in a thin line across the country like a highway; the underground city with nothing but parkland above it; the skyscraper city of Frank Lloyd Wright (a single towering building a mile high); floating cities; underwater cities; and many more.

Meantime on the outskirts of our cities there is no planning worthy of the name except that being done by developers. To help them cope with the problem of urban sprawl, and to save open space, an old idea has been revived—cluster development. Instead of dividing up the land in a featureless grid and placing each house in the center of its own plot of land, the builder clusters the houses, packing them back to back or side by side on a small portion of the site. This is a variation of making greater density to achieve more space. When only a small part of the land is used for the actual houses, the greater part of the site is left open and becomes a sort of commons belonging to all.

The cluster permits the builder to get as many houses on a tract as he would if he divided it up in the old way and subtracted individual roads, driveways and service areas. He can now leave the bulk of the land in its original state, preserving the features of the landscape instead of bulldozing it flat or filling it in. Trees, streams, a pond or marsh—all the features that brought people to the suburbs in the first place can be saved rather than destroyed. If the tract is large enough it can have a respectable piece of woodland or be laid out in trails, bridle paths, play areas for children, tennis courts, club houses and the like, all owned and managed in common. What the homeowner loses in strictly private ground he gains in a more usable public park whose ownership he shares.

Cluster developments do not appeal to the strongly individualistic, who recoil at sharing land and who prefer their own yard, no matter how stereotyped, preferably with a good, stout fence around it. But recently the cluster has been promoted in a new form as a condominium, offering a choice of apartments or town houses.

The usual arrangement is for the buildings to be attached in a line, using a minimum amount of land so that there is ample surrounding space. The purchaser pays for his apartment and there is an upkeep charge that includes grounds maintenance and various privileges—lake, swimming pool, clubhouse, and so on.

The appeal is to apartment-house dwellers, who, lacking a tradition of land ownership, see an opportunity to own their own dwellings, of a size suitable to them, and to share in the land and facilities without having to assume the entire burden of care. There is also an appeal to older people who want the open space but prefer to have the maintenance done for them. The one drawback has been the high cost of such condominiums.

Green Belt communities will be familiar to those who go back as far as the Roosevelt New Deal era, with their prototypes in Maryland and in Radburn, New Jersey. The Green Belt comes from an old concept, dating back to Queen Elizabeth and Oliver Cromwell. Elizabeth disliked and distrusted the growing metropolis of London and tried to contain it by ringing it with an agricultural Green Belt and forbidding the construction of buildings any farther—a constriction which didn't work.

In 1932 Sir Raymond Unwin proposed a variant of the Green Belt plan. Instead of a tight ring around London, he proposed a sort of broken girdle of scattered open areas that would have the effect of diluting the solid urban stretch with many contiguous park areas. His idea was not so much containment as to provide publicly owned spaces that would offer both eye relief and areas for recreation. Under the subsequent Green Belt Act of 1938 the British government bought or came to exercise control over 38,000 acres.

But by 1944, with London still expanding outward, the fear of uncontrolled sprawl was back, and a new plan by Sir Patrick Abercrombie reverted to the solid belt of containment. The new belt was to be five miles deep and was meant to separate the inner city and its suburbs from the rural areas outside. Further growth, accepted as inevitable, would be met by the establishment of entirely new towns outside the Green Belt. In 1955 the belt was deepened—to as much as 10 miles in some places.

In a sense it has worked. The Green Belt is there; it has preserved much of the lovely English countryside that would otherwise have been swallowed up. It has not worked in the sense of containing London, which has simply oozed under the belt and come up on the other side. The danger, as always, is that with continued growth —another million people in the London area are forecast by 1981— pressure will increase to sacrifice more parts of the Green Belt for housing.

Modern versions of the Green Belt are being actively furthered in other countries of Europe, where city planners have much more authority than in the United States.

Where topography permits, a favored European plan for a city is in the shape of a flower with slim, separated petals. The center of the flower corresponds to the dense inner city and each petal represents areas of suburban development along major highway and mass-transit lines. Between each petal is open space—parks, recreation areas, nature preserves, or farms. The net effect is a kind of irregular Green Belt that does not block further growth of the city in directed areas.

The success of this kind of planning is due to a very important element of control. Land use and transportation are combined in a single authority, thus the planners always know that transit lines will be available to service a growing area. And since they know where growth will occur, they cannot be confused by a sudden mushrooming appearance of apartment houses or shopping centers where no provisions for transit or services have been made.

The long wrangle in the United States over highways vis-à-vis mass transit has been settled in Europe, with the city planners deciding realistically that mass transit is the only possible answer in large cities. Subway systems are being expanded in Paris, Hamburg, Frankfurt, Munich, Vienna, Stockholm, Helsinki—even Amsterdam is pushing subway tunnels through its boggy underground.

Not that the motor car is being ignored in the new city planning for Europe, because ownership of cars is increasing very rapidly. Superhighways are being built and the new cities incorporate elaborate arrangements for bypasses and parking. Helsinki plans to put a large parking garage *under* a lake in the middle of the city, with access via tunnel. Thus the Finns will have parking without destroying the vista of open water.

But the Europeans know as well that to provide parking for everyone who wants to drive into town is both impossible and undesirable.

More and more streets are being closed off to vehicular traffic and converted into tree-lined, flower-decorated pedestrian malls. The celebrated Strøget in Copenhagen is a mile long and contains some of the city's best shops.

Parks are a must in the new planning. Amsterdam has projected its population to the year 2000 and is allotting 7.75 square yards per person in parks, beaches and recreation areas. Stockholm's plan calls for a small park in every neighborhood, with a sports field and a regional park large enough for "a sense of remoteness" from the city not more than 30 minutes away by car, bus or subway. If you visualize the flower-petal arrangement, with the regional parks between the petals, you can see that no part of the city would be far from a park.

The outward growth that overleaped the English Green Belt will not have the same effect in this new planning. Even though the city continues its outward expansion, the toothed design of the new Green Belt will retain the essential character of each of the neighborhoods and prevent their being engulfed in a shapeless, featureless megalopolis.

In the high-density center, European planners are tearing down old slums and rebuilding the heart of their cities. Glasgow is destroying five square miles of grimy tenements and replacing them with new apartment buildings, parks, pedestrian shopping malls and new transit lines.

Stockholm began a renewal program in 1962, which has drastically changed the business center, creating new streets, parking, stores, plazas and shopping malls and five new skyscrapers, which have not been greeted with unalloyed and universal joy by all the Swedes.

Planning for growth brought no serious problems in Stockholm because the city owns the land. It leases rather than sells land to builders, and maintains firm control over what is to be built. The city was never allowed the luxury of haphazard growth or sprawl.

This kind of planning in Stockholm goes back 330 years. The first master plan for Stockholm was adopted in 1640. In the last century a series of plans have been followed to keep abreast of a population that zoomed from 100,000 to 1,500,000 people.

Physically Stockholm is a beautiful city. It is built on a group of 20 islands in Lake Mälaren and Saltsjön—part of the Baltic Sea. The islands are linked by 42 bridges and connected by a modern subway system, which is comfortable, fast, quiet (ordinary conversation without screaming is entirely practicable while the train is running) and clean. Only a few stations in the center of the city are actually underground on most lines; the greater part of the trip is in the open or elevated, and as the trains cross the bridges they provide intriguing vistas of wooded islands and blue waterways.

There are few private houses in Stockholm proper. The population boom having proved louder than expected, the building of single-family homes was shelved in favor of apartments to provide the most living space on the least amount of land. The older apartment houses in Stockholm and its original suburbs are small—three or four stories high—resembling garden apartments in the United States. Great care was taken to preserve the landscape, and these small houses are often artfully snuggled into the terrain so that their brick or stone blends with trees and rocks and water in a setting least disturbed by man.

As early as 1904 the city assembly, anticipating outward growth, began buying farm and woodland outside the city limits. Some of this land was unused for 20 years but was acquired at prices that would seem ludicrous today.

Stockholm's master plan called for a central area of high density, with apartments surrounding the business area and tapering off to low-density housing farther out. A plan added in 1952 introduced the concept of the "new towns," or neighborhood units.

Prototypes were Vällingby and Farsta. Ironically enough, it is claimed that Vällingby was built on American principles, presumably meaning that it was designed around an American-style shopping center. The food supermarket is very popular in Sweden. Vällingby's central plaza also contains a youth center, cinema, library, schools and churches.

Two things are notable about Vällingby and Farsta. Both are small and both have visual appeal and charm. The central plaza, just off the subway, is attractively designed, and the neighborhood apartments are mostly of the garden type. The impression is one of open space and a parklike atmosphere. The subway ride from downtown Stockholm to Vällingby takes no more than half an hour and the train passes through other attractive suburbs and alongside a magnificent wooded park.

In contrast, the newest of the new towns near Stockholm, Skärholmen, is something of a jolt, having succumbed to the American credo that big is better. The subway opens onto a great plaza ringed with stores and featuring a huge supermarket. This is only the edge of a gigantic shopping center like any of the biggest in a prosperous American suburb—street after street of specialty stores. It is so overpoweringly large and contains so many stores that the local residents find it an irresistible fascination, and on Sunday families stroll the walkways just to window-shop the closed stores.

Worse yet, the apartment houses of Skärholmen march in identical and featureless rows like gravestones over the hills that surround the plaza. It is a curious hiatus in the usually excellent Swedish taste.

The Vällingby blueprint called for enough integral businesses and services to employ half of its projected residents. By 1960 there were 27,000 residents in Vällingby and 9,000 jobs. However, only 2,000 of these jobs were held by residents. The other 7,000 came from other communities. And 25,000 Vällingby workers commuted to other areas, half of them to Stockholm.

The new towns have many characteristics of an American suburb but offer more in the way of cultural and recreational unity, and have eliminated the inconvenience and the sterility of many suburbs in America. The goal is quality of life, which means different things to different people but is coming to mean integrity more than material prosperity. Few Swedes, for example, complain about taxes, which are considerably higher than American taxes. They know what they are getting for their taxes, but very few Americans do.

The apotheosis of new towns is now being planned in Sweden. Named Jarfalla, it will have about 100,000 residents and be located 12 miles from Stockholm. Gasoline-powered vehicles will not be allowed to enter Jarfalla. Electric minibuses will provide free transportation for passengers and baggage and will be supplemented by moving sidewalks. Heating elements under the sidewalks will melt snow as it falls so that removal will never become a problem. Garbage collection will be by pneumatic tubes from each house to an incinerator site 20 miles from the town. Heat and hot water, and presumably electric power, will be supplied by a nuclear plant. The air will be undefiled by smoke and the only sound in Jarfalla, according to its advance notices, will be the happy sounds of children at play. It is to be completed in 1976—for Swedish citizens only, unfortunately.

Can the new-town concept be applied here? Two new towns, Reston, near Washington, and Columbia, near Baltimore, have yet to prove out. One difference may be noted between Reston and, say, Tapiola in Finland. Reston devotes 56 percent of its land area to residences, 14.6 percent to industry and 14.5 percent to open space and recreational facilities. Tapiola devotes 24.2 percent of its area to residences, 5.5 percent to commercial enterprises and 55.9 percent to open space.

In the matter of open space the older American cities are more than niggardly. Compare Tapiola's nearly 60 percent with Detroit—5.5 percent; Pittsburgh—6.1 percent; Philadelphia—6.6 percent; Los Angeles—9.1 percent; Cleveland—14.8 percent; Chicago—20.5 percent; and New York—28 percent.

In *The Last Landscape** William Whyte says:

> As the major answer to the growth problems of the metropolis, the new town concept is not practical. For the New York metropolitan area the Regional Plan Association has figured that to take care of the expected population growth over the next two decades via the new town route, one hundred new towns would have to be built. This would take some doing and even if it were possible, the Association does not think it would be desirable. The result would be an extremely inefficient pattern, for the dispersal would rule out any effective mass transportation system.

The conclusion may be a little hasty. Housing for this new population will have to be provided, call it new towns, old towns, or any other kind of town. So long as it must be built, it would be far better to build it by plan than by a continuation of the urban sprawl we now have. If the European model of coordinating transportation with development were adopted, the problem of mass transit would be solved, along with the problems of open space and recreational potential.

Our neglect of mass transit in favor of the motor car has provided us with choking cities and clogged traffic. Characteristically, we are trying to cure ourselves with more of the disease—more highways to encourage more cars—instead of developing better mass-transit facilities.

* Doubleday and Company, Inc., 1968.

Most cities are widening their access roads and generally improving the gateways in the hope that this will speed up the snarled traffic. But this only brings in more cars, worsening the problem. Every effort to provide better roads merely deepens the flood of cars.

The remedy is exactly the contrary course of action. Every effort should be made to increase the inconvenience of driving a car into a major city—up to and including the outright banning of cars from the central business district. Such a ban would end congestion, the appalling waste of stalled traffic, and most of the pollution. It is so simple that it will take a new generation to do it.

CHAPTER

8

The Road Builders

THERE is a rumor that the highway lobby cherishes a dream to pave New Jersey over solidly and convert it to a 600-lane expressway.

It is a real tribute to the road builders' ambition, of which a Federal official once said, "They won't quit until the whole damned country is paved."

To a highway engineer it is obvious that New Jersey, the fourth-smallest state, is a natural corridor between the densely populated centers on the East Coast, so why not stop all the nonsense about a garden state and make it what God clearly intended it to be? The thought is not so extreme. Highway studies for New Jersey have already projected the "need" for 40 superhighway lanes within the next few years. A start has been made toward that 600-lane ideal.

With New Jersey already well paved and Long Island scored like a cube steak with highways all pointing at New York, the logical fate for Manhattan Island would be to make it the world's largest parking lot. Only John Lindsay, a latter-day King Canute, bars the way, with quaint notions of closing Central Park and dozens of busy streets to cars on weekends, with—who knows?—some wild dream of someday banning cars altogether.

In the contest between highways and people it should be clearly understood that people will not be permitted to take up space needed for cars. The major cities of the nation have already allocated 28 percent of their space for parking, and now Federal highway planners are talking of a post-Vietnam program to spread $7,000,000,000 worth of high-rise parking garages around the cities.

America's road-building program is the biggest public works project in history, dwarfing such puny achievements as the Egyptian pyramids or the Great Wall of China. It rests on that critical word "need." Do we need more roads? The answer to that is yes, if we are going to produce 41,000 cars a day, as predicted by Detroit for the last years of the seventies.

"If we are not careful," said General Omar Bradley, now chairman of the Bulova Watch Company, "we shall leave our children a legacy of billion dollar roads leading nowhere except to other congested places like those left behind."

Already we have built a mile of road for each square mile of the country—a staggering thought. According to the railroad lobby (whose sour-grapes remarks on the subject are apt to be carefully researched), this pavement occupies an area as big as six smallish states: Vermont, New Hampshire, Massachusetts, Rhode Island, Connecticut and Delaware—quite a lot of land to lose. Apparently it is only the beginning, since state highway departments have already projected plans for the period between 1973 and 1985 to spend $17,400,000,000 annually on new roads.

The American Automobile Association estimates that by 1985 there will be about 170,000,000 vehicles—cars, buses and trucks—on the road—60 percent more than we now have. America's love affair with the automobile is no fickle infatuation. Two national surveys made in 1967 made a statement and asked a question: "The auto pollutes the air, creates traffic, demolishes property and kills people. Is the contribution the auto makes to our way of life worth this?"

Four out of five people answered "yes," even in big cities where owning a car is a test of courage and obstinacy in the face of astronomical costs and appalling inconvenience. Eighty percent of American families now own at least one car, and 25 percent own two or more.

So new roads will be built. Car manufacturers are blandly aware of the schizoid nature of such protest as exists: few people even among those who complain really want to give up their cars. James M. Roche, chairman of General Motors, is quoted in *Fortune* Magazine as saying, "I don't feel it's a crisis. We've talked about congestion for a long time. Back in 1929 people said that no more cars would be sold because there weren't enough roads to handle them."

One way of dealing with a problem is to ignore it. In contrast, Henry Ford II is taking a long look at the role of business in society's ills, and, judging from three speeches delivered at Vanderbilt, Harvard and Yale, he didn't like what he saw.

> . . . people can see that their material possessions have been purchased at a high cost in environmental pollution—dirty air, dirty water, ugly landscape.
>
> Modern industrial society is based on the assumption that it is both possible and desirable to go on forever providing more and more goods for more and more people. Today that assumption is being seriously challenged.
>
> The industrial nations have come far enough down the road to affluence to recognize that more goods do not necessarily mean more happiness. They are also recognizing that more goods eventually mean more junk, and that the junk in the air, in the water and on land could make the earth unfit for human habitation before we reach the 21st century.

Until the sense of luxury and power the personal car conveys to its owner is canceled by its accumulating difficulties, the momentum of car building and road building will be impossible to slow. By 1980 traffic is expected to clog our major cities noticeably more than now—40 percent more in Pittsburgh, 50 percent in Boston, 90 percent in Detroit, and 100 percent in Los Angeles. Can Los Angeles take a doubling of its congestion?

Detroit expects vehicular traffic to move more slowly in spite of $3,000,000,000 to be spent on new highways and $1,100,000,000 for a rapid-transit system. The Detroit city planner was serious who said any driving in that town had better be done before 1975. The day forecast by some anonymous wit may in truth be approaching, the day when all traffic comes to an inevitable standstill and the only solution is to pour cement over the whole mess and start all over again.

Faced with a remorselessly rising tide of cars, city and state planners seem able only to chart new superhighways. While a huge network of roads was built in the twenties, highway construction did not become a major social or ecological force until the birth of the superhighway. The old roads were often only two-lane asphalt affairs, winding and twisting about the mountains, avoiding conflict with the rivers, and easing their way into towns, where they filled up to create giant traffic jams. But the modern expressway doesn't tiptoe through the countryside. It slashes through mountains, tears up stream beds, destroys a wildlife area, gobbles a park, and, in an increasing number of cases, cuts a town in two.

Road building is the biggest business in the country, supported by a Federal subsidy that pays the states 90 percent of the costs. Naturally it is bolstered by an impressive lobby composed of car manufacturers, tire manufacturers, petroleum producers, automobile clubs, road-building contractors, cement and asphalt manufacturers, building and teamsters' unions, and dozens of trade associations. It has even been suggested that the governors of the states should be included in this lobby since few of them are willing to turn their backs on a share of the $4,000,000,000 annual Federal highway aid. It is a handsome donation, considering that the Army Corps of Engineers gets only $1,200,000,000 for all the damage it can do, and the Bureau of Reclamation only about one-third of a billion.

The highway lobby is one of the most difficult to counter, because it is necessary to head it off not just in Washington but at 50 state capitols, where the routes are actually planned. A new highway is plotted by a district engineer working in almost complete obscurity, and by the time the public finds out where a road is going, it is almost too late to change it.

The Interstate Highway Program was organized by Congress in 1944. As a sample of some expert lobbying, it was given the name National System of Interstate and Defense Highways.

The rationale was to provide a network of roads to move military forces in case we were invaded by a hostile power—a slightly paranoiac idea but one that thwarted any opposition before it got started by making it seem unpatriotic. The fact that we are in a missile age and that such highways would be useless in the event of a nuclear war was ignored.

In 1956 the real funding of the Interstate Highway Program was achieved, with Congress stipulating that 41,000 miles of road were to be built by 1972 at a cost of about $60,000,000,000 (raised from an original estimate of $41,000,000,000), and that 90 percent of this amount was to be paid by the Federal government.

Unfortunately, Congress discovered in the summer of 1970 that the program was lagging and was not likely to be completed before 1978. Meantime inflation had seduced costs and the Federal Highway Administration had sent up a revised estimate of $70,000,000,000 to finish the job.

That estimate was two years old by the time Congress got around to it, and construction bids had jumped a record 9 percent over the year before. Realistically, Congress added $5,000,000,000 more to the tab, making the new estimate $75,000,000,000.

By this time the original $41,000,000,000 projected in 1956 was spent. For this the nation got 30,000 miles of superhighway out of the planned 41,000 miles.

What concerns Congress now is that the 1956 law, which funnels tax revenues from gasoline, tires, parts, and truck fees directly into the Highway Trust Fund (also called the Ever-Normal Trough), expires October 1, 1972, while the authority to spend the money continues until 1974. The decision will then have to be made either to continue the windfall for road builders or to start thinking about other forms of transportation. The present law stipulates that these revenues can be spent only for building roads.

The law was not unreasonable in 1956, when cars were proliferating and narrow roads were creating massive, stalled snakes of traffic. The immediate problem was to speed the flow. No one really visualized the prodigious growth of cars and highways that was to follow, and certainly no one talked of pollution. There was, in fact, relatively little pollution as late as fifteen years ago. Cars pumped the same combination of gases into the air, and there were electric utilities adding sulphur dioxide, and paper mills releasing acids into the streams, and steel mills dumping pickling liquors and picnickers adorning the landscape with cans and bottles. But there were so few of each that it hadn't begun to be noticeable.

In 1957 our gross national product, a rough index of our material wealth, was $453,000,000,000. In 1969 it had grown to $728,000,000,000. In 13 years it had increased by $300,000,000,000—which is very good from one standpoint, but even better if you realize that in the preceding 13 years it had increased only $100,000,000,000. And in the next 13 years it is expected to increase by $500,000,000,000.

So things are moving faster, like the population increase. From the time of Christ it took about 1,600 years for the population to double. Now it is doubling about every 30 years, and the figure soon will be down to every 20 years if things continue as they are going.

All this speeding up is exactly matched by car production and highway building. The lavish outpouring of funds for highways has naturally produced the illusion that it will go on forever. The road builders have estimated their funding needs for 1970-85 at $320,-000,000,000. This is only enough for about eight trips to the moon, but it led the Washington *Post* to comment: "320 billion is enough money for the government to buy all the railroads in the country, repair their roadbeds, fill all of their needs for new equipment, operate their passenger and commuter trains without charge to the riders for the next 15 years and still have a big kitty left over."

The length of the Interstate Highway system—41,000 miles—seems insignificant, but for two reasons its impact is out of all proportion to the actual mileage. One, most of its route is through new territory, and in many states there isn't much new territory that can stand being torn up. Two, the right of way needed is about ten times wider than old-style roads, which needed only about 33 feet.

The sense of dislocation that superhighways can create is severe. The landscape changes so rapidly that all sense of familiarity with a region can be lost after an absence of only a few months. This kind of strain has now pervaded all of our society. Alvin Toffler has dramatized the sense of transience, of change taking place faster than is desirable, in his book *Future Shock.** Describing the impact of change assaulting the nervous system faster than it can be assimilated, Toffler poses the question: Is our society heading for a nervous breakdown?

The highway program has contributed its share to this new kind of trauma. Where resistance to new roads has sprung up it has been motivated by resentment not over the cost but over dislocation and destruction. Interstate 95, the main eastern highway from Maine to Florida, has run into trouble in Baltimore, Washington, Georgia and South Carolina, with important sections blocked by citizens' revolts.

In Boston the John J. Fitzgerald Expressway, an incredible eight-

* Random House, 1970.

lane, multilayered behemoth, cut through the old town, dividing the historic waterfront from the rest of the city.

In Baltimore a rare park designed by distinguished landscape architect Frederick Law Olmsted was lost under the concrete of the Jones Falls Expressway.

The 16 lanes of the Dan Ryan Expressway stamped a concrete divider across Chicago.

In Nashville a loop of Interstate Highway 40 cuts the black ghetto in two, creating 50 instant dead-end streets and destroying 80 percent of all the Negro-owned businesses in the county.

A "fair price" for homes and businesses offered in state condemnation proceedings means nothing to these Negro businessmen; they have no place to go. Commercially zoned property in other black areas is tight, and discrimination in white areas shuts them out. So the road, cutting off nearly 100 square blocks, leveling about 650 houses, 27 apartments and several churches, creates a walled ghetto in which the residents are suddenly reduced from comparative affluence to poverty and hopelessness.

Highway 40 had been discussed for more than 10 years. There was one public hearing in May 1957. When the blacks of Nashville asked about it they were told that the suggested route was only preliminary and subject to change. The only change that occurred was that the state suddenly allocated $10,000,000 for land acquisition and engineering studies, signaling a full go-ahead.

A black group, organized as the I-40 Steering Committee, got legal help and fought all the way to the Supreme Court, arguing that there had been inadequate public hearings and that the chosen route was racially discriminatory. The Supreme Court declined to hear the case, and the state ignored the recommendation of its own New York engineering consultants, who proposed an alternate route to avoid the Negro business district.

The right-of-way chosen avoids Vanderbilt University and Peabody and Scaritt colleges, but chops up three Negro campuses: Fisk University, Tennessee A.&I., and Meharry Medical College. The one Negro park in north Nashville is damaged and the only library isolated.

Likely enough the discrimination here is involuntary. The road engineer has only one goal, to build at the lowest cost per mile. In a city the cheapest route takes him through the most inexpensive housing areas, which most often are Negro neighborhoods. Outside the city the cheapest route takes him through a park or wildlife refuge, which costs nothing at all. The engineer does not compute social costs.

In New York the Lower Manhattan Expressway had been fought over for nearly 40 years and had actually been shown on city maps since 1941. As envisioned by Robert Moses, master road builder, it would put an end to the insane traffic jams caused by half a million cars inching their way through the clogged and narrow streets, where trucks are double-parked and where it may take an hour to move a block.

The expressway would link up with the New Jersey highways on the west and the Long Island parkways on the east and make it easy to cross Manhattan or simply to reach one's office from the suburbs.

And therein lay the fly in the ointment. Clearly the expressway was not meant for the residents of lower Manhattan at all; it was meant to benefit the prosperous denizens of the bedroom suburbs who merely visit Manhattan to work or play. It would wipe out the living quarters of 2,000 families and destroy 800 businesses employing about 10,000 people. In both cases the victims would be predominantly Puerto Rican or black, poor, and under handicaps at finding either new living quarters or new jobs. One route proposed would have cut through historic Washington Square Park, the only play area for children in that part of New York.

The opponents of the expressway, chiefly people of New York's east side, Chinatown and Little Italy, do not have a lot of political muscle, yet for 30 years they fought off superior forces to stall the actual building of the road.

Blocking Robert Moses in his prime was no mean feat; but after Moses departed city politics, they took on others of the influentials. Governor Nelson Rockefeller was for the expressway, as was his brother David, president of the Chase Manhattan Bank. Harry Van Arsdale, labor chief, was for it, and so was Mayor Lindsay. Opposed to the road as a congressman from New York, he had shifted to lukewarm support of it, apparently on the basis of a change in plan calling for a sunken roadway rather than an elevated structure, which presumably would lessen air pollution.

Even so, a study of expected air pollution from the expressway conducted by the city's Air Pollution Control Department was kept secret until Councilman (now Congressman) Edward I. Koch applied pressure and got hold of a copy, which he had evaluated by the Scientists' Committee on Public Information. The committee reported that pollution could "cause the physical collapse of some people near the Expressway."

Finally, in July 1969, Mayor Lindsay, in a new reversal of position which *The New York Times* called "both a political act and a sign of the times," cancelled plans for the controversial roadway. On August 21, 1969, it was demapped.

About the time that the expressway first appeared on New York's plans, Robert Moses was retained by the city of New Orleans as a consultant on its growing traffic congestion. Moses outlined a plan to build an eight-lane elevated expressway to run between the famous French Quarter and the Mississippi River, within arm's length of Jackson Square and the Basilica, the heart of the Quarter.

The protests of local citizens and architects were brushed aside. The Louisiana State Highway Department ruled that the route of the expressway could not be changed, because otherwise it would have to go through railroad yards, warehouses and similar industrial installations, all obviously of more importance than a historical district. The ruling outraged John W. Lawrence of Tulane's School of Architecture. "We are not talking about an Expressway," he said, "we are talking about an act of barbarism."

The Vieux Carré was saved in the end, not by local action as such but by the current Federal Highway Administrator, Lowell J. Bridwell. An imaginative ex-newspaperman, Bridwell was once quoted as saying, "We build roads now to accommodate what's there and what the planners say the future will be. The planners are usually wrong."

In New Orleans, Bridwell decided that for an additional $7,000,-000—a small sum in highway building—the expressway could be diverted toward the river, brought down from its stilts to ground level, and a pedestrian plaza built over it, clearing a river view all the way to Jackson Square.

He offered New Orleans a package with strings—clear the river front of ugly warehouses and industrial slums, stop building motels in the Quarter, ban cars from some streets and make others throughways to speed traffic. The governor of Louisiana was told that the original elevated expressway would not be approved, and, since it was contingent upon a 90-10 deal in financing, that was pretty much the final word. It was probably the best deal ever forced down a city's throat.

In Cleveland an eight-lane expressway was planned to cut through Shaker Heights (one of the few roads headed for a prosperous neighborhood), with a massive interchange that would have destroyed the center of Shaker Lakes Park. The park land at least was cheap, and it dislocated no people or homes. Opposition in Cleveland blocked eight miles of highway—the last link in the interstate system in Ohio—forcing a new land-use study.

Fighting another Ohio road through the picturesque Olentangy River Valley, the Columbus *Citizen-Journal* commented acidly: "Civil engineers are expert road construction planners. They are not community planners. This freeway should be part of a broad community plan. It should include the saving of natural beauty as well as the building of a road."

In Philadelphia the Delaware Expressway projected as the lowest-cost plan a ten-lane road in an open cut along the riverfront. The resulting uproar brought in most of the city's business and political figures and finally won a compromise by which the road will be covered and landscaped, forming a pedestrian walk between Penn's Landing and Independence Mall.

Atlanta, where 54 percent of the downtown area is already given over to the motor car for driving or parking, is engaged in a brawl over an eight-lane inner-belt highway that will demolish Morningside, an old and dignified section of the city. In Cambridge, Massachusetts, faculty from Harvard and M.I.T. joined homeowners in a crusade against a loop of the Inner Belt highway that would knock out 1,200 homes. In Washington, D.C., Negro opposition to freeway building—"white men's highways through black men's bedrooms"—sparked a number of racial confrontations and added bitterness to the 1969 riots in the nation's capital.

In San Francisco aroused citizens discovered almost too late that the Embarcadero Freeway, a double-decked elevated monster, would have shut out major views of the bay. Forty civic groups set out to disprove the adage that you can't fight City Hall. In a series of embattled hearings they finally carried the city council by the close vote of 6 to 5. The freeway was stopped, literally in midair, and it stayed stopped through a new attempt to get it finished and a proposal to dig a tunnel under the Golden Gate Park. In so doing, San Francisco turned up its nose at $250,000,000 in Federal highway funds.

In Chicago a citizens' revolt against a crosstown expressway on stilts—"the ugliest road ever planned in the U.S.," 22 miles long and 140 feet wide—defeated the project by outspoken action, a march on City Hall. The Federal Highway Administration thought twice and offered a $2,300,000 grant for a design team of engineers and architects to produce a more acceptable plan.

Baltimore found itself impaled on the auto horns of a dilemma as traffic threatened to paralyze that city back in the forties. Baltimore looked for help and called in an eminent consultant—surprise—Robert Moses, road builder. Predictably, Moses recommended an expressway to slash through the city, through an area known as the Franklin-Mulberry district.

The prospect unnerved city officials, who vacillated for 20 years, unable either to drop the project or to go ahead with it. Meantime the threat alone was enough to tip Franklin-Mulberry into a downward spiral and deterioration to a slum. The city fathers poured $15,000,000 into studies, looking for different routes and designing programs for mass transit. While they fiddled, the Interstate System caught up with Baltimore. Three massive roads came in from various directions and began knocking at the gates.

The major threat was from Highway 10-D, designed to take a flood of cars through the city's center and cut in two a prosperous middle-class Negro neighborhood called Rosemont. It would have created a huge interchange in the downtown area as well, gobbling up 160 acres by itself. In addition, it called for a fourteen-lane bridge across the harbor, which would have ruined any attempt at redevelopment in that sector, and it would have carved up two historical neighborhoods called Federal Hill and Fells Point. It would have destroyed 4,000 homes, removed from the tax rolls $28,000,000 in land, and abolished 6,000 jobs.

The irony of the situation was, as a survey subsequently showed, that more than 40 percent of the traffic to be loosed into the city was not headed for Baltimore at all, but was merely passing through.

The situation illustrated perfectly the leitmotif running through the storm music over roads. If planning is left to the engineers they will invariably use a slide rule calibrated to figure costs and benefits to the highway *user*, and to the highway user alone. The rights of all others do not appear on that slide rule. The engineer is not a community planner and is oblivious to the larger community values. He does not see a now glaringly obvious fact, that the highways that created the cities are now destroying them.

A Baltimore architect named Archibald Rogers, already depressed over the Jones Falls Expressway, rallied the American Institute of Architects and organized an opposition. He won support from David Barton, head of the Baltimore Planning Commission, and from City Councilman William D. Schaefer, who took it upon himself to hold up condemnation ordinances for Highway 10-D. Rogers' idea was to give responsibility for planning to a design-concept team composed not only of engineers but of architects and sociologists as well.

He enlisted Nathaniel Owings, of the firm of Skidmore, Owings & Merrill, to head the team, and won a $4,800,000 contract for planning and engineering. This aroused the ire of the highway engineers, who resented "petunia planting aesthetes, bird watchers and do-gooders" telling them how to design a highway. They were mollified only when the Federal Highway Administration agreed to pay 90 percent of the planning costs, in which the engineers would share.

The concept team took off from the idealistic springboard of a notion that a road should not disfigure a city but make a contribution to its structure and beauty as well as serve its transportation needs. Its report came out in strong opposition to the proposed 10-D route.

The mere threat of cutting Rosemont in two had already polarized the city. "The whites will use the road to get to their high-paying jobs in the city," said Mrs. Esther Redd of the Relocation Action Movement, "while they live out in the country and pay low taxes. The poor will be pushed out of their houses."

The study also showed that the interchange planned for downtown would be overloaded by traffic, as would the new bridge over the harbor. A bypass was suggested instead, which the design team concluded would be preferable in every way.

The issue was deadlocked, with the concept team in one camp voting for 3-A, the bypass, and the engineers and some city officials holding out for 3-C, the original route. Mayor Thomas D'Alesandro III invoked his authority to make a decision. "If all of you," he said, meaning the engineers and officials, "are in favor of 3-C it's got to be wrong. I am adopting 3-A and I don't want to hear any more about the difficulties. I want to hear how it will be done."

His decision was made with the clear realization that the new route would be four and a half miles longer and cost $216,000,000 more. But a larger issue than money was faced—the issue of freeing a city from the tyranny of the automobile. Not to mention easing one cause of racial tension, and, for once, choosing human values over mechanical ones. Much more money has been squandered on less worthy causes.

The design concept caught on. In the threatened Franklin-Mulberry district of Baltimore, the new plan shows the expressway going underground, with shopping centers, schools and community facilities built over it. A new rapid-transit system is planned to run between the lanes of the highway. There are, it seems, ways to save even highways from themselves.

The effect of a highway on a city's ecology is one thing, its effect on open country is another. Since the road does not go through densely populated areas, consciences are even freer about destruction, the victims being merely landscape or wildlife. Roads have been thrust through several national wildlife refuges without much opposition from the Department of the Interior. The Wildlife Management Institute publishes an *Outdoor News Bulletin*, which recently commented on three refuges disturbed by interstate highways and added:

> Roadbuilders use the same arguments at every proposed refuge crossing. First they object to alternate routes that are suggested to spare wildlife values on the grounds of added construction costs. Next they compute a cost-to-the-public figure based on the theoretical expense to theoretical motorists who may travel the slightly longer stretch of highway. These figures ignore the fact that right-of-way across a wildlife refuge or other public land is free. They also fail to take into account damage done to wildlife refuge developments as well as the permanent loss of wildlife lands occupied by the highway.

In discussing the economics of highways, transportation engineers use a highly theoretical method of computation they call the "benefit-cost" ratio. The "benefit" is the expected savings to motorists in using the new highway. Motorists' driving costs are calculated on the length of the route, the grades, traffic estimates and so on, multiplied by the cost of operating the vehicle. This in turn depends on the type of car, driving speed, costs of gasoline, oil, tires, brake wear, and similar factors.

The "cost" end of the "benefit-cost" ratio is the cost of building the highway plus annual maintenance. The lowest building cost is assumed to be the most desirable unless an alternate plan, even if more expensive, will result in a lower user's cost.

Note again that the only people considered here are the highway users; the impact of the highway on all others, or on the environment, is not brought into any part of the equation. The effect on the surrounding community, the consequences of pollution, the destruction of scenery and wildlife—none of these are given any weight at all.

Dr. Dennis R. Neuzil, Assistant Professor of Civil Engineering at the University of Delaware, a member of the American Society of Civil Engineers and of the Institute of Traffic Engineers, believes that conservationists too often try to fight a highway planned for a scenic or wilderness area armed only with honest indignation. The engineers are vulnerable on their own grounds, and he describes a method in the *Sierra Club Bulletin* for January 1968 to show that their benefit-cost ratios are as leaky economically as they are often damaging to the ecology.

For a practical example Dr. Neuzil selected a proposed section of Interstate 70 in the Colorado Rockies about 75 miles west of Denver. The route is from Dillon westward to Vail, passing the Gore Range-Eagles Nest Primitive Area. Route 6 is already in existence; from Dillon it makes a wide loop south, then turns north and climbs up through Vail Pass at an elevation of 10,603 feet.

One proposal is to put in a new road parallel with Route 6 but wider and better, with an average grade of 3.1 percent and a design speed of 50 miles an hour. It would be 27.3 miles long and would cost $22,803,000.

The other proposal is to head straight west from Dillon instead of making the southward loop, and tunnel through Red Peak, a 13,000-foot mountain, at an elevation of 10,587 feet. This road, called the Red Buffalo route, would be nearly 11 miles shorter than the Vail Pass route, but its grades would be steeper, averaging 5.1 percent instead of 3.1 percent. It would cost $63,095,000, or nearly three times as much as the Vail Pass route, and would destroy about 7,000 acress of wilderness area in the Gore Range-Eagles Nest vicinity.

The Colorado Highway Department believes the costlier road to be justified on the grounds of savings to the users. They estimate the cost of driving this road at $8,457,000 annually as against $11,750,000 for the Vail Pass route.

To compute the benefit-cost ratio the department used a formula that yielded a saving of $1.94 to each motorist using the road. The break-even point is considered to be one dollar, therefore a theoretical saving of $1.94 makes the road economically justifiable.

However, says Dr. Neuzil, the formula used is open to question. There are three basic factors involved: interest rates, life of the road, and travel-time costs.

State highway funds come from the motorist, therefore highway improvements should create a saving to him equal to the interest he might have earned on that tax money if it had remained available to him for investment. What interest could he earn on an investment? The highway department used an interest rate of 3.5 percent, whereas in actuality it is entirely possible to earn 8 or 9 percent on high-grade bonds. The interest the motorist himself pays for financing his car is a very good guide, and this will run considerably in excess of 6 percent. Any interest rate of less than 6 percent is unrealistic.

If, then, the benefit-cost ratio for the Red Buffalo route is recalculated at an interest rate of 6 percent, the saving drops from $1.94 to $1.28. At 8 percent it drops to 99 cents and becomes uneconomical.

The second factor is the life of the road. A longer life spreads the construction cost for a lower annual cost, hence a better benefit-cost ratio. However, it is necessary to distinguish between the economically useful life of a highway and its physically useful life. A good highway may be structurally sound for 40 years, but it may become necessary to build a new road long before that time either to accommodate more traffic or a shift in traffic corridors away from the original route. If the old road thus declines in use before its costs are amortized, the investment has not paid off, and it is written off as a loss to the taxpayers.

The highway department's figures for the Red Buffalo route show a life of 20 years for the pavement, 40 years for the roadway, and 60 years for the tunnel and the right-of-way. Dr. Neuzil considers this optimistic. He rates the pavement at 15 years, the roadway and tunnel at 30 years. On this basis the benefit-cost ratio drops to 91 cents, considerably below the break-even point.

In any case, Dr. Neuzil believes that making traffic forecasts as much as 60 years in advance is unrealistic. Sixty years ago, for example, there was no highway traffic to speak of, and anyone who predicted the network of roads and cars that now exist would have been considered overimaginative. Within only a third of that span—20 years—the entire era of the electric car came and went. The interurban electric railways—which many people thought would eventually link all American cities—grew up, flourished briefly, and were slain by the motor car. For a few years you could almost cross the continent by trolley car, and now they are gone without a trace. Forecasts, says Dr. Neuzil, cannot be considered reliable for more than 20 years into the future—if that.

Finally, there is travel time, with two factors—passenger cars and trucks. In computing truck costs, the highway department assumed that trucks travel at the same rate of speed as passenger cars and that the operating cost of a truck is equivalent to a specified number of passenger cars. That number is usually considered to be six, but the Colorado Highway Department decided to use eight, giving them a decided edge in the figuring.

The engineers also ignored the fact that the Red Buffalo route would be much steeper than the Vail Pass route, although shorter, in using the same truck-to-car ratio for both. The result was to overestimate the saving to trucks in using the Red Buffalo route.

If the truck factor is then recalculated, with a 6 percent interest rate and a 30-year life instead of 60, the cost-benefit ratio drops to 90 cents.

In calculating travel time a value of $3 an hour is often used for heavy trucks, although Dr. Neuzil suggests that in some cases $5 an hour would be closer. Travel time for passenger cars is difficult to assess since so much of it is recreational and not commercial. Assuming that travel-time costs should be assessed for no more than half the cars on a rural highway, arbitrary figures were adopted by the Highway Department of $4.85 an hour for trucks and $1.55 an hour for passenger cars.

Assessing travel-time costs for all vehicles, the Highway Department reached its figure of $1.94 savings, based on an interest rate of 3.5 percent, which would drop to $1.10 at 6 percent. If full travel-time costs are allowed for trucks and half of the passenger cars, the saving drops to $1.47 at 3.5 percent interest and 83 cents at 6 percent interest. Finally, if no allowance is made for travel-time costs, the savings decline to $1.32 at 3.5 percent interest and 75 cents at 6 percent interest.

The lesson to be learned, Dr. Neuzil emphasizes, is that opponents of a highway need not be intimidated by the road builder's impressive documentation that a new highway is economically justified. The figures should be analyzed as to interest rates, useful life of the road, and travel-time costs before anything is conceded.

Of course nothing has been said so far of intangible values, such as scenic beauty, which are difficult to incorporate into benefit-cost ratios. Defenders of truth and beauty may still find themselves tongue-tied before hard-headed pragmatists like the highway official in Washington who confronted a delegation pleading to save the famous Beaverkill Creek in New York's Catskill Mountains. "What makes you believe," the man asked, "that a river is more important than a concrete highway?"

The record does not show that anyone asked him if he could build a river.

Incredibly enough, truth and beauty have sometimes won over concrete and benefit-cost ratios. Not often, but sometimes. A fifteen-mile segment of Interstate Highway 93 in New Hampshire has been held up—at this writing—for further study. The road, cutting through Franconia Notch in the White Mountains, would pass close to an ageless monument, a natural rock profile known to millions as the Old Man of the Mountain. Conservationists had objected that vibrations from traffic might dislodge rocks from the formation and destroy the world-famous landmark and tourist attraction.

Federal Highway Administrator Francis C. Turner announced the decision to halt the road at the sixty-eighth annual convention of the American Road Builders' Association in March 1970. It was the first time of any consequence, he said, that an interstate highway had been stopped from going through a park because of objections to its effect on the environment. Clearly he did not consider this a victory for his side, as somewhat ungraciously he called it a tempest in a teapot. "More rock falls off there from freezing and thawing in the winter than we're talking about."

Nevertheless, three months later Transportation Secretary John A. Volpe reaffirmed the decision that Interstate 93 would stop before going through Franconia Notch. A new interchange would be built from which motorists could reach scenic Kancamaugus Highway on the east or Routes 112 and 118 on the west.

State highway officials appeared anything but entranced by the deal, grumbling that New Hampshire's north country area would be cut off from major commerce and tourist arteries. Secretary Volpe said the decision had satisfied both conservationists and highway boosters "for the foreseeable future," but didn't know how long that would be.

So does even a state like New Hampshire, which has mostly scenery to sell, look the Federal gift horse in the mouth.

Another road stopped for the time being is the controversial second highway across the Great Smoky Mountain National Park, between North Carolina and Tennessee. In view of the tremendous traffic jams on Route 441, the only road crossing the park, the Park Service had proposed in 1966 a second transmountain crossing.

The opposition was vocal and loud. The Smokies are the finest remaining wild lands in the East. Moreover, since the Wilderness Act obligates the Secretary of the Interior to review every roadless area of 5,000 contiguous acres or more in the national parks for inclusion in the Wilderness System, parts of the Smokies were obviously in line for such inclusion. But not if any more roads were built.

While the fight over the transmountain road dragged on its usual way, two new roads were built: Interstate 40, northeast of the park, and a "Foothills Parkway" on its southeastern boundary. The completion of these roads drew off most of the through traffic. As a result Secretary Hickel in April 1970 ordered the Park Service to restudy its road proposal and report back in 1971—an order that hopefully tables the project indefinitely.

As though the Hudson River did not have troubles enough with mere pollution, despoilment of its banks and Consolidated Edison's determination to hollow out Storm King Mountain for a pumped storage plant, there was for a time an active campaign to build a six-lane expressway along a stretch of the eastern bank. The Hudson

River Expressway was to run from Tarrytown to Crotonville, a distance of 10.4 miles, and cost $125,000,000 to build.

One of the leading sponsors of the road was New York's Governor Nelson Rockefeller, who, speaking for other members of his family, offered to donate 130 acres of land along the route for a park near Mount Pleasant.

Not only conservationists were outraged by the proposed expressway. The Citizens' Committee for the Hudson Valley and the Village of Tarrytown joined forces in opposition. And questions were asked as to whom the new road would benefit. New York Congressman Richard Ottinger made no bones about his opinion. He said bluntly it would benefit the Rockefeller interests, which, he said, were bringing pressure to bear on Secretary Udall to approve the necessary permits for construction to begin.

Permits were necessary because the plans called for building part of the roadway on 10,000,000 cubic yards of fill dumped into the Hudson, extending in places as much as 1,300 feet out from the shore and drastically altering the shoreline. Since authority to dredge or fill any waterway is the responsibility of the Army Corps of Engineers, the engineers had obligingly issued a permit to the state, contingent upon approval by Secretary Udall. Udall, at first opposed to the expressway, finally said on December 11, 1968, that he would no longer object to issuing the permit.

A coalition of the Citizens' Committee, the Village of Tarrytown and the Sierra Club filed suit in Federal District Court. A temporary injunction against construction was denied by the court, but an early trial was set.

The trial began April 16, 1969, and ended June 11. Judge Thomas F. Murphy handed down his verdict on July 11, holding that the Army Corps of Engineers had exceeded its authority in issuing the permit without the consent of the Secretary of Transportation and Congress.

The decision was appealed, and on April 20, 1970, the United States Court of Appeals for the Second Circuit upheld the lower court ruling and continued the ban on the proposed highway. Some kind of record was set—the highway was one of the largest public works ever permanently enjoined in a lawsuit. The lawyer for the plaintiffs in both trials was David Sive of the Sierra Club.

As the trial ended, a spokesman for Governor Rockefeller said further appeals to the U.S. Supreme Court would be considered. But state lawyers pointed out that this wasn't really necessary. All that was required was to do what had been originally omitted—getting the necessary approval from Congress and the Secretary of Transportation.

The threat to the river, therefore, may come up again, and probably will. But to Dave Sive and other conservationists, a very important principle has been reaffirmed in this suit: recognition by the law that private citizens or groups have the right to sue in cases where they have no direct economic interest. It had been considered necessary for a plaintiff to show the threat of a direct financial loss as a result of the project he intended to stop.

The roster of highways aimed at parks or wildlife refuges is a long one and the story in each case is monotonously the same. The engineers point out that (a) the land is "undeveloped," therefore nobody will be hurt by taking it over, and (b) it is public, so it costs nothing. If there is a river or stream, so much the better, as (c) a stream bed offers a ready-made, easy route through rugged country.

Highway engineers like streams because the water has already cut a channel, promising easy grades. The stream bed also may be a supply of gravel for the builders. If the stream meanders, however, there could be the expense of additional bridges for crossing the curves. In that case the lowest-cost principle dictates a policy of digging a new straight course for the water—converting it in effect to a ditch—beside which the highway can run. The ditch, of course, is just that, no longer a river and no longer the habitat of such prized marine life as trout. "I cannot believe," said one Federal Highway administrator, "that the welfare of wildlife should be given a priority over that of our citizenry, nor the national defense effort in the area involved."

The statement contains only two unproved assumptions: that roads equal welfare, and that national defense is involved. What role roads would play in a missile war is hard to see.

In Montana a long war between the highway builders and the trout-stream defenders has been fought to a standstill. State conservation officials had kept a count of streams altered by highways and when the total reached 2,000 they reacted with alarm. A third of the state's rivers or creeks had been damaged and had lost 80 percent of their fish.

Senator Lee Metcalf of Montana took up the cudgels and launched a campaign called "S.O.S.: Save Our Streams," larding the *Congressional Record* with examples of mutilated rivers in his state.

John C. Peters, biologist of the Montana Fish and Game Department, tried to convince the highway engineers that there might be alternatives to their chosen methods. "It became painfully clear," he reported, "that they would listen but could not implement major proposals for minimizing damage." Put somewhat more simply, they would not change anything they were doing.

Citizens' groups began to form and helped draft a law, which managed to squeak through the legislature in 1963. It was limited originally to a trial period of two years, but it forced highway agencies to make their plans public at least 60 days before starting construction. This was little enough time, but it did give conservationists a chance to have a look and suggest changes at the last minute.

If the engineers rejected the changes suggested, the law required that the dispute go to a three-man arbitration panel composed of a representative from each of the warring factions and a nonpartisan third member to cast the deciding vote.

The trial period of this law was so successful in Montana that although it had passed by the barest majority it was renewed in 1965 almost unanimously. In the six years from 1963 to 1969 a total of 259 road projects were reviewed. The Fish and Game people asked for changes on only 88 projects. The arbitration panel never had to be called.

Road routes were changed to save the Madison, Big Hole, Missouri, and Blackfoot rivers. Where the route could not be changed, as in the St. Regis and Clark Fork rivers and Prickly Creek, special meanders were built to keep the channels as long as they were before the construction. Instead of straightening and shortening the Beaverhead and Missouri, additional bridges were built to preserve the natural meanders. Flood-plain vegetation removed during construction was replaced. What channel excavation was done was reserved for seasons when the trout were not spawning.

One result has been to save some money. As much as 15 percent of highway cost is design, and if the Fish and Game experts can see the design early enough, changes can be made before the beginning of construction makes alterations more expensive.

"The myth that the Stream Preservation Law would scuttle the road building program has vanished," Mr. Peters said (*New York Times*, March 26, 1970). "The law has shown the public that a construction agency and a conservation agency can work together, given the necessary legal framework."

Time will tell if his optimism is justified.

The damage that highways can do to the ecology of a wild area is for some reason not well documented, although it is hardly obscure.

Right at the onset, clearing and excavation for a right-of-way fells trees, tears up and removes the topsoil and disrupts the normal pattern of water distribution.

Cuts made through banks and hillsides start and encourage erosion. The channels washed out by rain are visible in practically every bank alongside a road. A fill in a low area often creates an artificial bank, offering the same prospect of erosion. In either case there is a steady sliding of soil and rock down onto the roadbed or into the area below the road. The drainage from these banks or filled areas, or even from the shoulders of the road, carries minerals free to move because they have been pulverized by digging and crushing. They wash out with the rain and filter into adjacent streams, changing their chemical balance.

Once the road is paved, the hard surface causes accelerated runoff, an erosion factor in itself. It decreases the surface area available for absorption of water and changes the direction as well as the rate of runoff, which further upsets the drainage patterns.

The minerals leached out by the rain are now joined by petrochemical compounds from car exhausts, plus lead and the familiar array of oxides, not to mention fine particles of rubber from tires and asbestos from brakes. The inevitable picnics along the way add food debris, paper, cans, bottles and all the varied elements of garbage. All of these find their way in some measure to the soil and water.

The road attracts cars, creates its own traffic. As the life of the road expands, it becomes a moving barrier to the motility of wildlife and a death trap for thousands of animals. A road is as divisive as a knife blade, cutting a wild area in two as though a fence had been put across it. It is a concrete ribbon of intrusion and disturbance *if* the original intent had been to keep a wild area in its natural state. If the function of a great park is to preserve an area of natural terrain, roads are incompatible with that goal.

There is a very fine line here between the need to make wild areas available to people and the need to keep them from being overrun by traffic and trampled to death. To put highways through parks is self-defeating. It abolishes the opportunity for isolation, it attracts huge crowds, which demand so much in the way of recreational and convenience facilities as to create an artificial environment totally unrelated to the natural conditions supposed to be maintained. The real danger is that in every such situation there are special interests always pushing to build another road (or another dam) for which there is no real need.

There is a problem, and it is linked to the American love affair with the motor car. Should the motorist be able to drive to any point he wants to see without getting out of his car? Must there be scenic loops to the brink of every canyon so the driver in a hurry can get a glimpse of the view through his windshield before he rushes on to make the next park before sunset? Or should he be inconvenienced to the point of being forced to walk a few hundred feet, so that a park may be kept a park rather than a parkway?

It is already late for us to make a choice. But there is, here and there, a tiny sign of encouragement. There is the suit filed by a Long Island group to block a Suffolk County highway on the once preposterous grounds that it will contaminate two wildlife preserves and kill fish and marsh birds. The group behind the suit call themselves the Yaphank-Middle Island Defenders of the Environment, and their lawyer is Victor J. Yannacone, Jr., of the Environmental Defense Fund, who shows no reluctance at all to take on the big ones.

More poignant, perhaps, is the action of California's Division of Highways, which is actually building a new swamp—a swamp for long-toed salamanders. It seems that the widening of a highway threatens to wipe out one of two swamps that are the remaining habitat of a little reptile left over from the Ice Age.

When a University of California zoologist, Dr. Robert Stebbins, called it to the attention of the highway division, the state came up with $3,000 to build a new murky, stagnant swamp stocked with all the necessities of life for salamanders.

Said Dr. Stebbins in a classic remark, "A creature doesn't have to be as big as a redwood or a whooping crane to receive protection."

CHAPTER

9

Man and Pesticides

FOR DDT and its relatives, the persistent insecticides of the chlorinated hydrocarbon types, the tide turned in 1969. Half a dozen states and several foreign countries had already restricted or banned their use altogether.

Federal action came in November when Secretary Robert H. Finch of Health, Education and Welfare ordered an end to the use of DDT in residential areas within 30 days of the order and outlined a phase-out program to ban all but "essential uses" by the end of 1971. The ruling included aldrin, dieldrin, endrin, heptachlor, chlordane, benzene hexachloride, lindane, and compounds containing arsenic, lead, or mercury. Other Federal agencies, said Secretary Finch, agreed in principle and would help HEW determine which uses could be considered truly essential.

The other agencies were the Departments of the Interior and of Agriculture, but principally Agriculture, Mr. Finch observing that the decision to impose the ban was harder for Secretary Clifford M. Hardin to make, with a constituency of farmers.

The ban was not total by any means. "It is not in the best interest of the public," said Secretary Finch, "to permit unduly precipitate or restrictive action based only on anxiety. The indiscriminate setting of zero tolerances for widely distributed pesticides could have a disastrous effect on our national supply of essential foods."

The statement is debatable. The boom in agricultural production came in the thirties, before the invention of the hard pesticides, and was due mostly to better seed, greater use of fertilizers, and improved farm machinery. Further gains since then have been relatively small.

There are safer, shorter-lived insecticides, but few are as cheap and convenient as DDT. Moreover, DDT's very persistence makes it attractive to many users, since it eliminates the need for constant reapplication.

A special HEW Commission on Pesticides had provided Secretary Finch with a double-barreled argument.

First, tests made at the National Cancer Institute showed that DDT produced tumors in laboratory mice fed a diet containing 140 ppm (parts per million). These tests were confirmed by similar work done at Roswell Park Memorial Institute of Buffalo, a respected cancer research laboratory. Among the test mice there was also an incidence of birth abnormalities and hereditary defects. Additionally, it was suggested that DDT may combine in a dangerous way with other chemicals normally harmless—a contingency suggested in 1962 by Rachel Carson in *Silent Spring,** for which she was berated as "unscientific."

Second, the persistent pesticides are stored in body fat and retain their dangerous qualities for a long time—as much as 10 years, Secretary Finch said, which was an understatement. The half-life of DDT is at least 15 years—a period in which the potency of the material is reduced by half.

The Federal action failed to satisfy at least two groups of people. One was the pesticide manufacturers, six of whom appealed the decision and indicated they would fight it in court. The other was composed of environmentalists who filed suit in the U.S. Court of Appeals to demand a stricter ban on the grounds that the HEW action was full of loopholes.

This suit was brought by four conservation groups and six individuals, including the Environmental Defense Fund, the Sierra Club, the West Michigan Environmental Action Council and the National Audubon Society.

One of two petitions filed said that the action on DDT to control shade-tree, house and garden, and aquatic pests did not prohibit

* Houghton Mifflin Company, 1962.

the sale of the chemical, but only required a change in the wording of the labels.

The second petition, filed by the Environmental Defense Fund and six California residents, asked for a review of an earlier petition on foods that had been rejected by HEW. That petition had asked for zero residues, and the department had rejected it on the grounds that DDT was now so widely distributed in nature that even if all use were discontinued immediately, it would be many years before it disappeared from agricultural products. The Environmental Defense Fund replied it did not expect zero residues immediately, but would like to see a legal ban, with waivers that would allow diminishing traces over the next several years.

Federal action had come reluctantly, lagging behind Michigan, Maryland, California, Arizona, Florida, Wisconsin and New Hampshire, all of which had placed varying restrictions or outright bans on chlorinated hydrocarbon pesticides. Early in 1970 Governor Rockefeller of New York had asked the state legislature for a bill to prohibit "all unnecessary uses of persistent pesticides such as DDT in New York State."

Outside of this country, Canada, Australia, Sweden, Denmark and West Germany had outlawed or restricted the use of DDT. The German action was taken in spite of the fact that only 47 grams per hectare (2.47 acres) were used in that country compared with 261 in Sweden, 514 in the United States, and 615 in Japan. The only Western nations still manufacturing DDT remained Italy and France.

The ground swell against DDT has resulted in a number of spontaneous revolts, such as the one in 1966 against the Suffolk County (Long Island) Mosquito Control Commission, spearheaded by a vigorous, ecology-trained lawyer, Victor J. Yannacone, Jr. That action was the genesis of the Environmental Defense Fund, which also enlisted Dr. Charles F. Wurster, assistant professor at the State University of New York at Stony Brook, and Dr. George M. Woodwell, ecologist of the Brookhaven National Laboratory. A third founder was Dr. H. Lewis Batts, Jr., director of the Kalamazoo Nature Center. The new organization promptly took on the Michigan Department of Agriculture in an attempt to stop the spraying of dieldrin over 2,800 acres of watershed draining into Lake Michigan. That case, lost on a technicality, was a trial run for EDF.

The charge against DDT and its relatives is that they are not just pesticides but biocides—literally, killers. They kill mice and bats, fish and birds. At least one exterminator uses DDT lavishly for killing mice, remarking ingenuously that if it were not for DDT he would have to use poisons. And while defenders of DDT insist there is no evidence of harm to human beings, recent reports are more than straws in the wind.

Dr. William B. Deichmann of the University of Miami School of Medicine has found that patients with high blood pressure, leukemia, liver cancer or other carcinomas had three times the residue of DDT or similar pesticides in their body tissues as persons who had died accidental deaths.

Soviet scientists have reported that persons exposed to larger than normal quantities of DDT in their work show disturbances of stomach and liver function.

Dr. Irma West of the California Department of Public Health testified that in 1965 one California farm worker died of pesticide poisoning and that between 100 and 300 others had been nonfatally poisoned. Another thousand had experienced dermatitis, chemical burns of the skin and eyes, and miscellaneous conditions traceable to contact with pesticides.

Two research scientists of the University of Colorado Medical Center in Denver, Dr. J. H. Holmes and Dr. D. R. Metcalf, did a study on 150 men working in organophosphate plants. The study included electroencephalograms and sensory evoked responses, psychological testing, physical and neurological examination, and intensive interviews.

The scientists found that men repeatedly exposed to these chemicals were literally slowed down. They had lost energy, were more irritable, had poor coordination and demonstrated greater memory loss. They had difficulty walking a straight line, picking up and handling small objects, and tracking accurately with their eyes.

On twelve of the men severely exposed the doctors recorded electroencephalograms during their sleep. Nine had narcoleptic sleep patterns (brief periods of stupor) and two had excessive REM (rapid eye movements) characteristic of narcoleptic persons.

Dr. Metcalf has suggested, from his observations of these men, that organophosphates work on nerve centers in the midbrain, inhibiting cholinesterase, an enzyme involved in the transmission of nerve impulses and in the mechanics of sleeping and waking.

The peculiar danger of DDT is due not only to the fact that it is one of the most active substances known but also to a factor called biological amplification. DDT is not soluble in water but combines readily with fat. It is stored in the fatty tissues of animals and there accumulates. Thus a trace amount in the soil is concentrated in earthworms and further concentrated in the robin, which eats many worms. Anyone who has ever seen a robin spinning dizzily on a lawn, shaking with uncontrollable tremors until it dies, has had a firsthand demonstration of the effect of DDT on the nervous system.

This combination of storage in fatty tissues and long life makes meaningless the official levels for residues. For example, the merest trace of DDT in the ocean—one part per billion or less—might be considered harmless. But the traces are absorbed by trillions of diatoms, the microscopic single-celled plants at the base of the food chain. Diatoms are ingested by larger animal plankton in huge numbers; if one rotifer eats a thousand diatoms, the DDT accumulating in its tissues will have been concentrated a thousand times. The next-larger animal up the food chain—shrimp or various larvae or small fry of a fish species that feed on animal plankton—escalates the level another thousand times, and so on.

By the time it gets to the bigger fish and the fish-eating birds, the DDT or its major metabolite, DDE, has reached a viciously toxic level. In their appeal to Governor Reagan to halt the use of DDT in California, marine biologists put the DDT amplification from the bottom to the top of the food chain at 100,000 times.

The bottom mud in parts of Lake Michigan contain .014 ppm of DDT. The small crustacea living there have concentrated it to .41 ppm. Fish, feeding on the crustacea magnify the level up to 6 ppm. And the herring gulls of the lake, eating fish, build up the level to 99 ppm, which is enough to kill them when food is scarce and they begin to burn their own fat, releasing the DDT.

The coho salmon introduced into the Great Lakes after the lampreys and alewives were checked, excited fishermen until the FDA found DDT residues as high as 20 ppm. Commercial catches were seized and sale of Michigan coho forbidden.

Filter-feeding animals in the sea may concentrate pesticides even further. An oyster or clam doesn't merely eat the small tidbits that come its way; it draws in and filters water through itself constantly to strain out small edible creatures. This constant filtering processes a much larger quantity of water and exposes the animal to a greater volume of insecticide. And oysters live in shallow water, where the pesticide contamination is heaviest. These filter feeders achieve much higher concentrations of pesticide—half again as much for a clam as for a mud snail. Oysters have been found with 70,000 ppm of chlorinated hydrocarbons in their tissues.

Compare this with the tolerances for DDT residues on food established by the FDA: one to three parts per million on most foods, seven parts per million in the fat of meats. Then consider the fact that a farmer often uses more than one chemical to spray a crop. As biologist Clarence Cottam suggests, if that healthy-looking, unblemished, worm-free vegetable you buy contains only half the legal limit of each of four different chlorinated hydrocarbons—for example, DDT, dieldrin, chlordane and lindane—the effect is like getting twice the legal limit of any single one. When laboratory rats were fed a diet containing the "harmless" level of two parts per million each of DDT, lindane, toxaphene, chlordane and methoxychlor, liver damage resulted.

The first warning on DDT came as early as 1949, when it began to show up in milk. Today, however, cow's milk is safer than human milk, which carries two to four times the quantity of DDT. One British study found the DDT concentration in human milk 32 times higher than in cow's milk. The FDA has set a residue level for milk at .05 ppm and has paid farmers more than $1,000,000 to dump milk that exceeded this level of DDT.

A disturbing discovery about DDT is that it has shown up in the fat of penguins at the South Pole and in remote Antarctic seas thousands of miles from the scene of any spray program. It has been learned that DDT is carried by wind and water over the globe in very much the same patterns as radioactive fallout.

Dr. George Woodwell explains the spread of DDT in *Scientific American* (March 1967):

This seems to be true in spite of the fact that DDT is not injected high into the atmosphere by an explosion. When DDT is sprayed in the air, some fraction of it is picked up by air currents, circulated through the lower troposphere and deposited on the ground by rainfall. Migrating birds and fish can transport it thousands of miles. So also do oceanic currents. DDT has only a low solubility in water but as algae and other organisms in the water absorb the substance in fats where it is highly soluble, they make room for more DDT to be dissolved in the water. Accordingly, water that never contains more than a trace of DDT can continuously transfer it from deposits on the bottom to organisms.

Describing the persistence and accumulation of DDT, Dr. Woodwell says:

In a Maine forest we found that where spraying had been discontinued in 1958, the DDT content of the soil increased from half a pound per acre to 1.8 pounds per acre in the three years from 1958 to 1961. Apparently the DDT residues were carried to the ground very slowly on foliage and decayed very little. The conclusion is that DDT has a long half-life in the trees and soil of a forest, certainly in the tens of years. Sampling along the south shore of Long Island, that had been sprayed for 20 years to control mosquitoes, disclosed concentrations in the mud that ranged up to 32 pounds per acre.

The unique action of DDT is one of interfering with enzymes normally secreted by the liver. In birds this liver malfunction cuts the amount of estrogen in the system and inhibits calcium metabolism. The bird lays eggs with shells so thin that they collapse when incubated; or in some cases with no shells at all.

America's national symbol, the bald eagle, is rapidly disappearing through its inability to reproduce. The peregrine falcon, which once nested on the Palisades of the Hudson—and, indeed, on some New York skyscrapers, where it subsisted on pigeons—has become extinct in the East and scarce elsewhere. A report on the American osprey from Rhode Island shows that the birds had declined from 200 pairs in 1940 to six pairs in 1969. The osprey is a fish eater, and fish are prime magnifiers of DDT. The Everglades kite is on the verge of extinction, and the number of pelicans is dropping sharply. The brown pelican is Louisiana's state bird, but no pelicans are left in Louisiana—or probably in Texas either.

In addition to the fish eaters, raptorial birds, such as hawks and eagles, are prime victims of biological amplification, because they are at the top of the food chain, eating animals that have already concentrated large amounts of DDT in their fat. Apart from a certain sentimental attachment to the eagle as an American symbol, there is no widespread affection for hawks or other predators. Farmers shoot hawks despite their protected status because they believe them to be chicken thieves—although it would be difficult to get the idea of ownership across to a hawk. But the major food of hawks has always been rodents, and they perform an indispensable service in keeping the rodent population in balance. Other birds of this type are carrion eaters and are equally important in maintaining an ecological balance. Eliminating these birds would create problems of imbalance, and the population explosion among rodents would undoubtedly lead to an uncontrolled poison campaign, with further disastrous effects.

In all that has been said so far the danger to man appears circumstantial rather than direct. For instance, the Chester Beatty Research Institute in London reports that the persistent pesticides act as alkylating agents in the body and could double the mutation rate as well as increase the carcinogenic activity. But there is a real and direct threat to man (and other animal life) in the increasing contamination of the ocean with DDT, which is endangering the vital oxygen supply.

While a large part of the earth's oxygen is produced by land plants, an even larger proportion—perhaps 70 percent—comes from the tiny one-celled plants in the sea, the phytoplankton. These plants—diatoms and desmids—are green plants containing chlorophyll, as does a leaf, with the same function of converting water and carbon dioxide into food by sun energy, and giving off oxygen as a by-product.

DDT in trace amounts, only a few parts per billion, has been found to reduce photosynthesis by as much as 50 percent. At a time when man's demand for oxygen is rising, when carbon dioxide and pollutants from internal-combustion engines are pouring into the atmosphere, when open land is being paved at the rate of 1,000,000 acres a year, the loss of oxygen from the sea could be catastrophic. And even apart from the reduction of oxygen, the effects of DDT on the phytoplankton must inevitably damage the bottom rungs of the food ladder and reduce the amount of life in the sea.

While the defenders of persistent pesticides have become less audible as the evidence accumulates, they are not entirely stilled. As late as December 1969 the World Health Organization and the Food and Agricultural Organization said the use of DDT is essential to assure sufficient supplies of food, particularly in the developing nations. They acknowledged that there are "still some unresolved questions about the effect on the environment and on human health, but agreed that DDT and some other organochlorine insecticides will continue for some time to have a vital role to play in food production and crop protection in many countries." They conceded, however, that the use of DDT should be restricted to those pest problems for which there are no other satisfactory solutions and that "all unnecessary and excessive use" be avoided. ("Under no circumstances," says Dr. Woodwell, "can DDT be controlled out of doors.")

A report the same month from the official British government Advisory Committee on Poisonous Substances Use in Agriculture and Food Storage paralleled the WHO report. It recommended that some uses of DDT and other pesticides be curbed, but on the whole found no evidence of adverse effects on man and said there was no case at present for their withdrawal.

An attack on conservationists more reminiscent of the old days came from *Barron's*, a financial weekly, on May 5, 1969. The arguments are virtually a text of those used for years. Headlined "Up with People and Down with the Venomous Foes of Chemical Pesticides," it charged: "In liberal folklore, where reform is a magic word, and business and greed are synonymous, *Silent Spring* and its message—that modern man, through use of chemical poisons, is despoiling his environment and threatening the balance of nature—have become an article of faith. Others are not so sure."

The article labels as an unsubstantiated scare the imminent extinction of the bald eagle (which competent Audubon ornithologists have verified) and lands heavily on a statement from *Silent Spring* that vast acreages have been contaminated by pesticides. "In fact," refutes *Barron's*, "insecticides have touched less than 5% of all U.S. acreage and less than one-half of 1% of land hospitable to wildlife."

Just what "touched" means is not clear, though it probably means land actually sprayed. But Dr. Woodwell, speaking at a Rockefeller University–New York Botanical Gardens symposium, said that 1,000,000,000 pounds of DDT and its derivatives are now circulating through the world's air and water. Considering the persistence and the patterns of spread, it is unlikely that *any* part of the globe is untouched by DDT. In the last decade about 200,000,000 pounds of DDT have been used *annually*, worldwide.

In the remote Brooks Range of Alaska the hatching of young peregrine falcons has dropped from 2.5 young per pair in 1960 to one per pair in 1968. The unhatched eggs carry high concentrations of pesticide—although no one has sprayed the Brooks Range.

In the light of events since *Silent Spring* was published in 1962, the statements made by Rachel Carson seem restrained to the point of diffidence.

The *Barron's* article comments on the seizure of coho salmon from Lake Michigan and quotes Dr. Wayland J. Hayes, former chief of toxicology of the United States Public Health Service: "You can eat Coho salmon containing 19 parts per million of DDT morning, noon and night as your total diet for at least 19 years without any harmful effects."

Dr. Hayes might have checked with John Carr of the U. S. Bureau of Commercial Fisheries. Mr. Carr began testing fish for pesticides in 1965. He found that, as might be expected, the fat carried a heavier concentration of DDT than other parts of the fish. He found DDT levels in the fat of coho salmon as high as 105 ppm.

The Food and Drug Administration finally set the tolerance for DDT in commercially sold fish at five ppm, which automatically ruled out Great Lakes fish and threatened Michigan fishermen alone with a loss of $2,500,000 a year.

Barron's looked briefly at the "eggshell phenomenon" and decided it was for the birds.

> The Delta region of the lower Mississippi is one of the heaviest pesticide-use areas in the United States, yet favorable quail populations have been maintained. In a similar vein there are abundant pheasant populations in the aldrin-treated cornfields of Iowa. Tests at the Patuxent Research Center (U.S. Bureau of Sport Fisheries and Wildlife) has disclosed that birds (mallard ducks, bobwhite and coturnix quail, pheasants and cowbirds) can not only survive but show normal weight gain on a (DDT) level five times higher than is likely to be encountered in the natural environment.

The choice of the Mississippi as an example was perhaps unfortunate since it has been the scene of massive fish kills due to runoff of pesticides from farms. But there are other things wrong with the argument.

For one thing, graminivorous pheasants and quail cannot be compared with carnivores at the top of a food chain, where they receive the full amplified accumulation of stored DDT. As to the health of the birds, it has been amply demonstrated that fish and birds with high levels of DDT in their bodies show no overt signs of ill health, but the effect on the liver enzymes has a direct bearing on their reproduction.

At Patuxent the experiments performed on mallards were described by Lucille F. Stickel during the Wisconsin pesticide hearings. Mallards fed 3 ppm of DDE laid eggs with shells 13.5 percent thinner than the controls, which would break six times as often and produced less than half as many ducklings.

Sparrow hawks fed 1.4 ppm of DDT and .28 ppm of dieldrin laid eggs with shells 15 percent thinner than controls, with a proportionate reduction in fledglings—59 to 61 percent, as against 84 percent for the controls.

Switching to India, *Barron's* pointed out that malaria has been nearly wiped out by the use of DDT. The claim is more optimistic than true. The anopheles mosquito has become completely resistant to DDT and malaria is again on the rise. Eradication is just about as far away as ever. *The Wall Street Journal* reported on April 14, 1970: "The U.S. alone, for example, has contributed about $400 million since 1956 to the worldwide fight against malaria. Yet after dramatic early gains, the disease is on the increase again. Malaria and two other scourges of under-developed countries, trachoma and schistosomiasis, now infect 800 million people—almost four times the U.S. population, almost one-fourth the population of the world."

In India, Pakistan, Iran, Iraq and Mexico, says the paper, some mosquitoes have developed resistance not only to DDT but also to the insecticides replacing it. Because of the limitations of DDT, medical teams in El Salvador are switching to drugs to prevent malaria. Still, cases have risen in this small country from 15,000 annually to about 83,000 in 1967.

Each time malaria seems eliminated from a locale, it comes back from a reservoir of untreated individuals. In Ceylon there were only 17 cases reported a few years ago, out of a population of 13,000,000. But from 1968 to 1970 there were 4,000,000 cases.

Finally, we have the gypsy moth. Said *Barron's*, "Two impressionable New England states which stopped using DDT a few years ago, suffered heavy damage from the gypsy moth."

The Audubon Society sees it differently.

In our opinion this is an overrated forest pest. Officials in Pennsylvania, New Jersey and New York are hysterical over it just now, but southern

New England states which have lived with it have learned to relax. Occasional heavy defoliation may kill, at the worst, one in 20 trees that would not have lived anyway, most often on ridgetops (or sandy areas) that produce nothing valuable as timber. This culling process actually benefits most northeastern woodlands, which are overstocked with oak. What is needed—and what the U.S. Forest Service has at last undertaken —is basic research on the ecology of gypsy moth-forest interaction.

The ironic thing about DDT is its increasing failure to control insects in the face of its increasing danger to fish, birds and mammals. Insects that were bowled over by DDT soon metabolized it, then actually became addicted to it. Their resistance to DDT occurred for a very simple reason—their shorter life span.

A human generation is something over 20 years. Other animals range downward to months or weeks. And, like man, only a few animal offspring are produced at one time.

Insects have reproductive cycles of weeks or even days, with each female laying hundreds or thousands of eggs at a time. Evolutionary changes in insects, therefore, can occur quite rapidly, compared with man or other animals.

If a wholesale spraying of insecticide kills 99 percent of an insect population, the remaining 1 percent survives because, for one reason or another, it has somewhat superior resistance. This 1 percent will now produce a generation with perhaps 2 percent of resistant offspring, so that on the next spraying a few more will survive. The process goes on, doubling or better in succeeding generations until the entire species is resistant.

The first warning of DDT's failure came, in fact, when it was discovered that the common housefly had become immune. Since then at least 150 insects have learned to ignore or enjoy DDT. Scientists have bred houseflies in the laboratory that are not even made uncomfortable by a thorough drenching in concentrated solutions. The failure of DDT has led to the development of even more toxic hydrocarbons, such as dieldrin or heptachlor.

It was the discovery of DDT in milk, about 1949, that first hinted there might be a problem. Until then it had been a glamour product, credited, truthfully, with saving 10,000,000 lives by eliminating the mosquitoes that carried malaria and the body lice that spread typhus.

DDT—1,1,1-trichloro-2,2-bis(p-chlorophenyl)ethane—was first compounded by a German chemist named Othmar Zeidler in 1874, in an era when chemists were more interested in learning the mechanics of organic reactions than in developing marketable commodities. All Zeidler did was to make a note of its basic properties and put it on the shelf, where it remained for 65 years.

In 1939 a Swiss chemist named Paul Mueller was looking for a product. The Colorado potato beetle had arrived in Switzerland and was having a field day with the crop. Mueller was looking for a new insecticide. He tested hundreds of chemicals and came upon Zeidler's forgotten compound. It worked. It paralyzed the insect's nervous system in short order, a tiny amount was devastating, and it killed no humans.

The U.S. Army discovered it next. DDT went along on the invasion of Europe in World War II. Coming into war-exhausted lice-ridden villages, Army medics enthusiastically dusted the inhabitants with DDT and performed a medical miracle by eliminating typhus. Sprayed over marshes, DDT played havoc with mosquitoes. Everyone was delighted with it, predicted the end of insect-borne diseases, and in 1948 Mueller received the Nobel Prize.

Twenty-two years later Sweden couldn't take back the prize, but did it symbolically by banning DDT.

Actually, DDT would be quite the boon it was thought to be if it were not for its persistence. Misgivings about it were imparted by biologists and conservationists to a few alert people—some in government. President Kennedy showed awareness of a problem with DDT as early as 1961, referring to one government agency that encouraged the use of an insecticide that might harm birds whose preservation is encouraged by another agency.

That same year John V. Lindsay, congressman from the Silk Stocking district of New York, wrote a letter to the Secretary of Agriculture, asking about the pollution of the environment and inquiring if legislation was needed. He received a reply from the Assistant Secretary informing him that all was well and that no new legislation was needed.

The motives behind all this are interesting. In the first place, why do we need insecticides? Why are insects a problem? The major reason—in agriculture—can be given in one word: monoculture.

Nature produces a variety of plants, rarely an unbroken stretch of just one kind. In such a mixed field the insect's food supply is limited and scattered, natural predators abundant. Even when man began to introduce farms, the early ones were small. If insects made inroads, the farmer rotated his crop as a defense. The potato beetle found slim pickings in a cornfield and cabbage worms went hungry in a field of beans. So insects were kept in reasonable check and nature's balance of zero population growth was, for the most part, well enough maintained.

Then farms got bigger and more specialized. Now there were square miles of corn in Indiana, miles of cotton in the South, miles of wheat in Kansas. The corn borer found itself living amidst plenty undreamed of, the boll weevil had cotton to spare. Their numbers increased so enormously that the only answer was wholesale spraying with insecticides.

Enters now the resistance factor. The more a farmer sprays, the more he needs to spray, as succeeding generations of insects acquire resistance. He finds an "insect rebound," for he is not only breeding immune species, he is killing off the natural predators that helped keep them in check. So he goes to more and more toxic sprays or uses combinations of chemicals. And now, as the nondegradable materials accumulate in soil and water, the damage begins to show up in foods and milk and the tissues of animals.

What worries most ecologists is the thought that it is already too late, that forces have been set in motion that are irreversible and that, in the next 20 or 25 years, can destroy us.

Dr. Russel Earnest heads a Fish Pesticide Laboratory in Tiburon, California. He is conducting research to determine the level at which a pesticide will kill a fish outright, and the sublethal, or chronic, level at which it merely interferes with reproduction. The sublethal level, says Dr. Earnest, is the crucial one. He has found pesticides in the tiniest plankton and the biggest whales, in fish that never come near the surface of the water, and in birds that live on the open ocean without the slightest contact with man.

DDT interferes with fish reproduction in a different way than with birds. It accumulates in the yolk sac of the egg and kills the young fry after it hatches.

Kenneth J. Macek of Interior's Fish-Pesticide Research Laboratory at Columbia, Missouri, added DDT to the diet of brook trout at the rate of one milligram per kilogram of body weight—or a gram of DDT a week to a million grams of fish. This dosage did not kill adult fish but caused increased mortality among the fry. The amplification of DDT residues was comparable to that found in fish in lakes, including the Lake Michigan coho salmon. At two milligrams per kilogram, the mortality increased to 88 percent, compared with 1.2 percent in the controls.

Native trout may become a memory in New York's mountain lakes. The most celebrated of all fish are failing to reproduce in Lake George, and the State Health Department has issued warnings against eating trout found in a dozen northern lakes because of the high DDT residues. Whatever fish are left in Lake Erie are obviously not edible; the governor of Ohio banned all fishing in April of 1970.

The killing of millions of fish (and many ducks) in the Rhine when an insecticide called endosulfan contaminated the river led to a number of complications. It is not known whether the insecticide was accidentally dumped into the river or was merely the runoff from a dusting program of the vineyards and fruit trees around St. Goar. Downstream, the Dutch were angry over what they considered to be insufficient warning from German authorities that the poison was coming their way. Reservoirs using purified Rhine water were hastily closed off and only emergency supplies of water were used.

French conservationists charged that this and similar insecticides were responsible for human deaths as well as those of fish. Antoine Reille, French conservationist, said that gulls in the Loire Valley were laying eggs without shells, that half the heron population of Britain, plus half the surviving 50 pairs of peregrine falcons in France were sterile. Angrily denying official German statements that endosulfan was harmless to humans, he said that an anti-mosquito campaign in southern France with a similar chemical had resulted in at least 75 cases of human poisonings, four of them fatal.

The official tolerances for pesticides on food set by the FDA offer no real protection, because FDA has never had enough inspectors to enforce its regulations. Even though responsibility is shared by the Consumer and Marketing Service of the Agriculture Department, checking is still a matter of spot sampling. For example, the 1969 crop of chickens and turkeys was 2,300,000,000 birds, and the eight regional CMS laboratories checked about 3,000 poultry carcasses for pesticides.

The complete inadequacy of the system was made embarrassingly clear when the Campbell Soup Company called the attention of the Agriculture Department to a shipment of turkeys it had received, all bearing the stamp: "Inspected for wholesomeness by the U.S. Department of Agriculture." The turkeys were heavily contaminated with heptachlor epoxide (1,4,5,6,7,8,8-heptachloro-2,3-epoxy-2,3,3a,4,7,7a-hexahydro-4,7-methanoindene), an insecticide three to five times as powerful as DDT.

The U.S. Department of Agriculture requires that killed poultry or poultry products containing one-half part or more of heptachlor epoxide per million be destroyed. Most of these turkeys contained one to three parts per million, but some had 17 parts per million—34 times the maximum allowance.

The turkey episode was hushed up for three weeks while the Agriculture Department investigated. It found that a group of turkey growers in Arkansas had been dusting their turkey ranges with heptachlor to kill chiggers, in spite of warnings on the pesticide containers that the chemical could harm livestock. Asked why he'd paid no attention to the warnings, one grower is reported to have answered, "Hell, I ain't raising livestock, I'm raising turkeys."

Campbell evidently had only a slim faith in government testing procedures and maintained its own residue-monitoring system to pick up chemicals in poultry and other products used in soups, frozen dinners and other packaged foods.

The story was kept under wraps until a newspaperman, Arthur E. Rowse, learned of it and asked for an explanation. A press statement was then issued—four weeks after Campbell had notified CMS that a large number of Arkansas turkeys (final estimate was about 350,000) were grossly contaminated.

Another episode never publicly admitted by CMS occurred in June 1969. Some 630,000 hens past the egg-laying stage were shipped to a Delaware food-processing plant to be converted into 57,000 pounds of chicken spread and broth. Some others were used by a New York State food plant.

At some point it was discovered that the fowl were contaminated with excessively high levels of dieldrin (1,2,3,4,10,10-hexachloro-6,7-epoxy-1,4,4a,5,6,7,8,8a-octahydro-1,4-endo, exo-5,8-dimethanonaphthalene). The Agriculture Department was able to trace the shipments, and it appears that none of this contaminated meat ever reached retail stores.

It was also not generally known that the Department of Agriculture was dusting airports liberally with dieldrin. The purpose of this program was to nip in the bud any foreign insect that had hitched a ride on an airplane arriving from overseas before it could set foot on American soil. Dieldrin was chosen, said Donald R. Shepherd of the department's plant-pest-control division, because one pound was equal to about eight pounds of DDT, and it would last five to seven years, minimizing reapplication.

Dusting the ground was pretty silly, commented Dr. Charles F. Wurster, Professor of Biological Sciences, State University of New York, "All the insects have to do is get off the plane and fly. The solution is to treat the airplane, not poison the environment around it."

Between 1954 and 1969 at least 23 military bases and 33 civilian airports were treated with dieldrin, and some were retreated. At least 75,000 acres were involved, and more than 250,000 pounds of dieldrin used.

Protests from conservationists halted a redusting of National Airport in Washington in October 1969. Tests showed that the dieldrin used in 1962 was still quite potent. The protesters made the point that dieldrin runoff could easily pollute the Potomac River, and Agriculture's Donald Shepherd admitted that there had been "a few fish kills" during the program. In Norfolk, Virginia, where applications had been intensive, he said state agencies had found dieldrin in oysters and shrimps, but "that it wasn't considered hazardous."

In July 1969 Agriculture's program to dust Kelly Air Force Base, 15 miles from downtown San Antonio, with 69,000 pounds of dieldrin brought Michigan's Congressman John D. Dingell into the act with a protest to Agriculture, Interior, and the Air Force. The measure, he said, "seems well in excess of any logical need—20 to 40 times the load they need to do the job."

"It's like scalping a man to get rid of dandruff," said Dr. Clarence Cottam. "The base is near the San Antonio River and only 150 miles from the Gulf of Mexico. That amount of dieldrin, if it got away, would be enough to sterilize the bays all along the Gulf Coast."

There wasn't any danger, said Ned Moritz, the local Agriculture man in charge. There was a chance "we'd kill some fish."

After a story on airport dusting appeared in the Washington *Star*, the Agriculture Department announced it would switch from dieldrin to chlordane, the "softest" of the hard pesticides but one considerably more toxic than DDT.

Malathion is a much safer insecticide than the persistent chemicals. It is not very toxic, does not accumulate in fatty tissue, and is broken down by natural processes after a few days of exposure. Yet even malathion poses problems in use—it kills honey bees and fish.

A typical confrontation took place in South Carolina in 1966. Under a $490,000 grant from the U.S. Public Health Service, the state Board of Health undertook a mass spraying program to eradicate "target mosquitoes"—potential carriers of yellow fever and dengue. At least so said officials of the Board of Health, although there had not been a case of yellow fever in South Carolina for more than 30 years.

Spraying began in July with malathion, diluted to a strength of two ounces to an acre. On July 13, Jules W. Lindau IV, a beekeeper of Columbia, sought an injunction in Richland County Civil Court on the grounds that his bees were being killed. The injunction was denied by Judge John A. Mason, who transferred the action to Harry M. Lightsey, master in equity.

The spraying continued, and in a few days Wilbur L. Gunter, a fish-pond owner, complained that after the spraying he had collected three garbage cans full of dead trim and catfish. Mr. Lightsey then ordered the spraying stopped until the entire matter had been heard and the court had rendered a decision.

On Long Island, target of extensive spraying for mosquito control, suburbanites who left their cars outside found that malathion dissolved the paint—a result more disconcerting than toxic.

A more serious program was the campaign against the fire ant, an import from South America, which spread through a number of southern and southwestern states. Fire ants were accused of ruining the ranges and inflicting extremely painful bites. Whether they were really a menace or not is a moot point; it depends upon who is telling it.

The Department of Agriculture decided it was a menace and in 1957 unveiled a campaign of propaganda and pesticide. Great areas were sprayed with dieldrin and heptachlor—two of the most toxic and persistent of the chlorinated hydrocarbons—resulting in a massive kill of wildlife, but without much effect on the fire ant. So in 1970 Agriculture unwrapped a new weapon in its continuing warfare—microencapsulated insecticide.

Grains of dechlorane are coated to create tiny capsules, which are then broadcast over the fields by airplane. About 10,000,000 acres are being treated of the estimated 38,000,000 harboring fire ants.

The theory is that the ants will pick up the little capsules and bring them back to their nests, where they will crack the shells and eat the poison. The Agriculture Department says the encapsulated pesticide retains its toxic potential for 30 days, compared with 48 hours for the same insecticide broadcast in powder form. Also, a smaller dose is required, perhaps only one-fifth as much. Who or what else may eat the capsules besides fire ants is not specified.

Insecticides are not the only headache of irresponsible technology. For nine years U.S. military planes sprayed South Vietnam with chemical defoliants, principally 2,4,5-T (2,4,5-trichlorophenoxyacetate)—called "Orange" in military jargon. And for the same length of time American officials assured the public and the South Vietnamese that the spray was harmless to animals and humans, in the face of accumulating evidence that the chemical induced fetal deformities in mice and rats, and may contribute to the occurrence of cancer.

A study begun in 1966 by the Bionetics Research Laboratories for the National Cancer Institute indicated gross birth defects in laboratory mice, according to preliminary findings revealed in February 1969.

Ned D. Bayley, Agriculture's director of science and education, said that the deformities found by Bionetics Laboratories may have been due to a contaminant in 2,4,5-T called dioxin, which has occurred in concentrations as high as 27 parts per million. However, FDA studies showed birth deformities in mice occurring with dioxin in concentrations of less than one part per million.

In October 1969 Dr. Lee A. DuBridge, science adviser to President Nixon, announced that spraying with 2,4,5-T in Vietnam would be confined to areas remote from population. Further evidence on the safety of the defoliant, he said, was being sought.

There had been some evidence out of Vietnam that received

very little attention. On December 1, 1968, an American C-123 took off from the Bienhoa air base, northeast of Saigon, and shortly developed engine trouble. The pilot jettisoned 1,000 gallons of defoliant over the villages of Tanhiep and Binhtri in 30 seconds.

An Air Force medical team visited Binhtri shortly thereafter and reported that the villagers had suffered no ill effects. No doctors came to Tanhiep at all, and the stories told by some of the inhabitants, who said they had become dizzy, vomited and had to stay in bed for several days, received no attention. Reports of later miscarriages and birth defects were ascribed to bad diet and the hardships of war.

Meanwhile, back at the ranch, about 3,000,000 acres of U.S. rangeland were being sprayed each year with 2,4,5-T to kill brush and trees and create more pastureland. In 1968 the Department of Agriculture paid farmers $8,800,000 for brush and weed control, mostly by the use of 2,4,5-T.

In November 1969, following his October announcement on Vietnam, Dr. DuBridge announced that the Department of Agriculture would outlaw the use of 2,4,5-T on food crops if the FDA did not set tolerances for residues. The spraying of pastureland to clear brush would be continued.

Then, on April 15, 1970, three departments of the government—Agriculture, Interior and HEW—acted in concert to suspend the registration of 2,4,5-T, making interstate sales illegal. In addition the Defense Department announced that spraying in Vietnam would be halted—temporarily.

Since not only farmers but home gardeners had been using 2,4,5-T for weed control, the government issued a warning that there was "imminent danger" to women of child-bearing years from liquid 2,4,5-T. The warning cautioned gardeners about disposing of unwanted stocks, since flushing the chemical down the drain, burying it, or burning it merely released it into the environment in a more concentrated form than spraying.

The registration ban did not affect the sale of this herbicide already in garden-supply stores, nor did it prevent gardeners from continuing to use the stock they already had on hand. And none of the actions interfered in any way with the thousands of gallons of 2,4,5-T still drenching pastures and cattle ranges, roadsides and wooded areas.

In 1965 forest rangers working on an irrigation project in the Tonto National Forest in Arizona began a spraying program with a combination of chemicals that included 2,4-D, 2,4,5-T and Silvex.

There are small farms in this national forest area east of Phoenix, among them one belonging to Robert McKusick, who is a potter and who also raises goats. Until 1965, McKusick said (*New York Times,* February 8, 1970), of 200 kids born to his goats only one was dead and none deformed. After the spraying program started, he estimated that 60 percent of his goats have been born dead or deformed, and his chickens have all but stopped reproducing.

On Richard Lewis' farm were 12 peach and apricot trees he had babied for 25 years, hauling water to keep them thriving in that arid climate. "Within days after the rangers sprayed last summer," he said, "the fruit on the trees shriveled into hard black lumps." More serious, several members of his family who came out to watch the spraying have since been afflicted with chest pains, swollen feet and a variety of respiratory problems.

James Andrews was exposed to the spray as he helped a friend build a shed near the border of the national forest. Eight months later he had lost 25 pounds and 30 working days. He had severe chest pains, but an electrocardiogram showed no heart malfunction. He went to Los Angeles to consult a specialist, who made extensive tests, which allegedly revealed herbicide poisoning.

Other reports from the area say that miscarriages have increased, women have suffered internal hemorrhaging and a number have had hysterectomies. Robert Courtney, supervisor of the Tonto National Forest, suspended the spray program but denied that the spray was responsible for any illnesses. Reports issued by state and Federal inspectors said the complainants had "blown up a rather minor happening into staggering proportions."

"The real danger," said Robert McKusick, "is not to us or our animals, but to the environment. We can always move in or out, but the environment is here to stay. Go up to the sprayed area. There aren't any birds, and there won't be any."

The pattern has become appallingly familiar. Our technology, resourceful as it is, too often is a one-step affair. We have produced new and potent substances. What are the long-range effects? We don't know, but let's try them and see. It is an attitude parallel to that of the engineers who want to dam a river. Will it harm the ecology of the area? Let's try it and see. The fact that the damage may be irreversible doesn't alarm them.

The pharmaceutical industry, which has been testing new drugs for many years, has a very simple rule. If a new compound is dangerously toxic to test animals it is abandoned without further ado. Teratology studies are routinely performed, and if a drug induces foetal abnormalities it is dropped. Under those guide rules neither the persistent insecticides nor the herbicides would ever have reached the market.

But if we are to feed our large and growing population, we are committed to monoculture. And if we have monoculture, with its encouragement of insects, we must have some form of insect control. What can be done, outside of poisons?

Apparently a good deal can be done. In Florida a plague of screwworm flies that killed cattle were effectively controlled without insecticides. Batches of male flies were sterilized by cobalt-60 irradiation and released to mate with female flies. No offspring resulted from these matings and the fly population plummeted.

A great deal of work has been done, and is being done, primarily by government and academic researchers, to find new ways of controlling insects through the use of materials that are not toxic in themselves and that are selective, so that they do not kill foe and friend alike. Such methods include sterilization, the use of natural predators or natural parasites of insects, and hormones.

One of the most promising materials is the juvenile hormone, discovered by Dr. Carroll Williams of Harvard. Insects go through three stages in their development—larva, pupa and adult—and hormones control the growth and metamorphosis from one stage to another. The juvenile hormone is synthesized by the corpora allata, a pair of glands in the head.

The intriguing thing about juvenile hormone is that the larva needs it to develop normally, but it must then stop being secreted if the mature larva is to metamorphose into its pupa stage. Later, after the adult has been formed, the juvenile hormone is again needed. An additional finding of Dr. Williams was that insect eggs will not develop normally in the presence of juvenile hormone.

The potential of juvenile hormone as an insecticide was recognized as far back as 1955, when the substance was first extracted from the glands of a male *Cecropia* moth. An application of the hormone to the larva simply stops it from ever developing into its next stage. An application to eggs prevents them from hatching out into larvae.

The hormone has irresistible advantages: it affects no other forms of life but insects; it is not toxic (even to the insects themselves, except in the sense that it scrambles the blueprints of development); and it does not seem likely that insects can muster an immunity to it.

Extracting enough of the natural hormone from the tiny *Cecropia* moth glands was out of the question, and the next step was to synthesize it. Several preparations were made. One, by John H. Law, made from farnesenic acid, was about 1,000 times more active than the crude oil first milked from the *Cecropia* moth.

A team of chemists, under William S. Bowers of the Department of Agriculture's Beltsville Laboratory, produced a synthetic analogue of the hormone (10,11-epoxy farnesenic acid methyl ester) that resembled it structurally and was about .02 percent as active as purified *Cecropia* hormone.

Identification of the molecular structure was accomplished by a team at the University of Wisconsin headed by Herbert Röller. The chemical structure is $C_{18}H_{36}O_2$. The pure hormone is so active that one gram can kill a billion insects. It is as effective against the body louse as DDT was originally, and even prevents the eggs from hatching.

At Beltsville, in 1967, other researchers reported two more experimental compounds: 22,25-diazacholesterol, and triparanol, which interfere with cholesterol assimilation in the tobacco hornworm.

At the University of Wisconsin entomologists found that vanillin and syringaldehyde, both extracted from lignin, attracted bark beetles, including those that carry Dutch elm disease, and so might be used to lure and trap these insects.

Other Department of Agriculture entomologists have experimented with encapsulated bacteria, planting them in the ground to control the European corn borer.

A cross-breeding experiment with mosquitoes, mating a strain native to California (but bred in Germany) to Burmese females (*Culex pipiens*), resulted in eggs that did not hatch and virtually wiped out the mosquitoes in the test area.

Dr. George B. Craig, Jr., of the University of Notre Dame, reported on a study in which he used the seminal fluid of a dozen different species of male mosquitoes to effectively sterilize female mosquitoes for a period of 10 weeks—the entire mosquito mating season.

Dr. Williams, the discoverer of juvenile hormone, working with the Czech biologist Karel Sláma, found that beetles were being killed by the paper towels used in their laboratory. The towels were made from balsam fir pulpwood, and the active ingredient was isolated and identified as the methyl ester of a substituted cyclohexenecarboxylic acid, chemically related to a material with juvenile hormonal activity known in Czechoslovakia.

The Department of Agriculture imported from India a wasp that finds scale insects tasty, in particular a species causing wide damage in pastures along the Gulf Coast. The wasp was grown and released by airplane over 900,000 acres in Texas, where it was reported scale-insect infestations were reduced by 50 percent.

At Oak Ridge National Laboratory, under a program jointly sponsored by the Atomic Energy Commission and the National Institutes of Health, selected viruses are being developed to strike at single species of insects. The virus must be carefully tested to be certain it is specific for one species only and harmless to plants, fish, animals and man.

Natural viruses have been used before—in Canada and Scandinavia, for example—but the Oak Ridge program concentrates and purifies the virus by centrifuging out all extraneous material from the culture in which they are raised, using newly developed zonal centrifuges. The purified virus is so concentrated that a spoonful will cover 1,000 acres. Moreover, it remains in the environment, dormant until picked up again by another insect, in which it multiplies, so that a single application may be good for as long as 25 years. Present cost is estimated at $3 an acre, but is likely to be reduced to 50 cents. Field trials are under way against the caterpillar of the tussock moth, the gypsy moth, the European pine sawfly and the bollworm.

Until now chemical companies have been relatively indifferent to the development of biological controls for insects. A report published by graduate students of the Harvard Business School * evoked so little interest that the authors commented: "Firms having vested

* "Selective Insect Control," by Philip E. Asquith, et al., published by Management Reports, Boston, Mass.

interests in traditional methods of control have little incentive to do research to obsolete present products (unless these products are endangered by other factors); too, firms not already in the field, with no vested interests, have no current overwhelming incentives to do basic research in fields promising very uncertain payback."

The Harvard group recognized that these companies had problems—high costs in meeting FDA and Department of Agriculture standards on safety and efficacy (although one wouldn't think so, considering the record of the chlorinated hydrocarbons and herbicides), the problem of obtaining exclusive licenses for products discovered wholly or partly in government laboratories, and restricted markets for selective insecticides.

Some work, however, has been done commercially. Four companies are marketing a bacterial toxin that is not dangerous to birds and mammals. The toxin, developed from *Bacillus thuringiensis,* is inexpensive, and its market, until now small, may be considerably stimulated by the ban on DDT. For town officials it appears to obviate the problem of poisoning cats, dogs and birds when shade trees are sprayed. Farmers who have used it like the idea of being able to spray close to harvest time without leaving a toxic residue.

Another company, which has developed a sterilizing agent for flies, boll weevils and other insects, has been discouraged by slow action on approval by FDA, with numerous requests for additional data. A similar complaint is voiced by the discoverers of a virus, *Heliothis zea,* isolated from cotton bollworms as long ago as 1961.

There seems little doubt that a nontoxic insecticide is within reach. What has not been solved is the larger problem of one-step technology. We can kill—very efficiently. But how much do we know of the aftereffects on our ecology?

"The decision to kill is always a serious one," says Roland C. Clement, Audubon biologist, "especially since most insects are beneficial." Echoing the thought is Dr. Carroll Williams, who wrote in *Scientific American* (July 1967) that all these materials "leave unsolved, however, the problem of discriminating between the .1 per cent of insects that qualify as pests and the 99.9 per cent that are helpful or innocuous. Therefore, any reckless use of the materials on a large scale could constitute an ecological disaster of the first rank."

Conservationists, by their Cassandra-like prognosticating woe, have, in fact, functioned as a brake on those enthusiasts always willing to try new materials on a large scale. This pattern, too, is sickeningly familiar. A new and potent substance is used. Biologists and conservationists in the field find the first signs of damage and utter warnings. Government officials respond with assurances that there is no danger and all is well. Then the damage becomes unmistakable and officialdom hastily backtracks and issues bans.

Conservationists have learned that protest and publicity, valuable as they are—indispensable in the long run—take effect too slowly to prevent much needless damage being done. For quicker action they have turned to the courts. "People need advocates," says Victor Yannacone. Government has its own advocates, industry has its own, political organizations have their own.

And Dr. Clarence C. Gordon, professor of botany at the University of Montana, told the American Trial Lawyers Association: "For a long time I made speeches, I thought that was the answer. Now I am convinced that the answer is in the courtroom."

When the Department of Agriculture admits, says Yannacone, that a pesticide, if checked at all, is checked only for its effectiveness on the target insect and not on beneficial insects or wildlife, this incredible lack of concern for the safety of the American people can be met only one way—knock on the courthouse door.

It is a knock that will be sounding more and more in the near future.

PART V

In the Public Domain

CHAPTER

10

Public Lands and Public Parks

WE seem now to have a paradox. Open space is vanishing, yet our big population centers occupy only 1 percent of the land—we are told by urbanologists. One percent doesn't sound like very much.

But perhaps it depends upon your definition of large population centers. On the East Coast, megalopolis is a reality from Boston to Norfolk, in places a hundred miles wide and with signs that it will expand westward to Chicago. Another monster is growing on the West Coast between Los Angeles and San Francisco, a third in the Southeast, a fourth in Florida, and there are developing massive sprawls in the heartland of the continent and in the Southwest. It looks like a lot more than 1 percent.

You can see this quite well from the air, but even closer contact with ecological disaster can be had from a car window. Drive anywhere and see havoc. Drive on a new eight-lane superhighway and see the bulldozers already at work on the shoulders, toppling trees, ripping up the earth to add four more lanes to the road. Where there were curves, the hills are being sliced through to straighten the road and save a fraction of a second for the impatient driver.

And as you drive observe the scattered junkyard of the American landscape—the blatant billboards, motels, gas stations, roadside zoos, used-car lots, frozen custard-hamburger-hot dog stands, freight yards, auto graveyards, factories, souvenir stands, diners, amusement parks, row houses, real-estate signs—the thousand manifestations of American culture. In the East particularly, the countryside has all but vanished under this onslaught; the only oases are the parks. And now that realization has come that more parks are needed, land is appropriately expensive and budgets are threadbare.

If cost is the factor, and government always poverty-stricken except for military spending, it is pertinent to ask the question: Who owns the land? Who is the biggest landowner in America?

Brace yourself for a surprise. The biggest landowner in America is the Federal government. You and me. We own more than a third of the nation's land—755,400,000 acres. Nearly one-half of all the land west of the Rockies and almost *all* of Alaska.

The bulk of it—470,400,000 acres—is controlled by the Bureau of Land Management in the Department of the Interior. Agriculture has nearly 187,000,000 acres, and the rest comes under the Park Service, the Department of Defense and other agencies.

Once upon a time nearly all of the United States was Federal domain—about 1,800,000,000 acres out of the two billion there are. Over the past 175 years the government has made strenuous efforts to sell it or to give it away, with such indifferent success that nearly half is still left, in spite of some determined land rushes as the tide of empire flowed west. Obviously, what is left is land nobody wanted very badly, if you except lands withdrawn for the national parks, wildlife refuges or Indian reservations.

So if the government still owns vast areas and we need more parks, why not just declare it parkland? It won't cost anything. But it isn't that simple. It never is, although recently the Bureau of Land Management classified 150,000,000 acres for Federal retention, 40,000,000 acres for further reclassification and 3,000,000 acres for disposal.

The story of how all this land came under Federal ownership is intriguing, as hindsight is always intriguing in the face of current dilemmas. It began with the American Revolution.

As the rebellious colonies demonstrated an unsuspected obstinacy and the war dragged on, King George of England felt the first qualms about losing his American empire. Following the surrender of Quebec in 1763 he proclaimed a line down the crest of the Alleghenies as the western boundary of the original thirteen colonies. This was the extent of the original land grants, he said, and from there west to the Mississippi the country remained the property of the Crown. Beyond the Mississippi was Spanish, clear to the Continental Divide.

The American colonies, however, had colonies of their own, and these included the Northwest Territories, plus Tennessee, Mississippi and Alabama. Daniel Boone had ignored the king's line as early as 1769 by establishing a settlement in Kentucky, and he was followed by John Sevier, James Robertson, Richard Henderson and others.

With the surrender of Cornwallis in 1781 the American republic became a reality. The thirteen states that now ratified the Articles of Confederation could no longer act as independent nations, so ceded their colonies, some 233,000,000 acres, to the new Federal government, blithely ignoring the objections of England, Spain and France. This was the nucleus of the public domain.

It was a brave gesture for the time, and a somewhat risky one, for it had little legal standing in the face of a solid opposition from three powerful nations. The success of the entire maneuver hinged on the outcome of some remarkable horse trading on the parts of Benjamin Franklin and John Jay at the peace talks in Paris.

Congress had appointed five men to make peace with England—Franklin and Jay, plus Henry Laurens, John Adams, and Thomas Jefferson. But Jefferson never got to Europe at all, and Laurens and Adams were delayed so long that Franklin did most of the spadework, being joined later in Paris by Jay. Acting on their own, and conveniently forgetting to clear their decisions with America's ally, France, they carried on the bargaining with the English negotiators.

Jay offered England a right-of-way up the Mississippi to Canada in return for a clear title to the Northwest Territories. It was an imaginative offer in view of the fact that no one really knew where the Mississippi was. Jay used a map published in 1755 by an old crony of Franklin's which showed the headwaters of the Mississippi in Canada; but it was off by a state or two. The Mississippi actually starts in Minnesota, and there is no way to sail from the Gulf of Mexico to Canada on its waters.

But the peace treaty was signed, and the Territory of the United States northwest of the Ohio River came into being and was ratified by the Northwest Ordinance of 1787. It officially added to the new nation the land that now comprises Ohio, Indiana, Illinois, Michigan and a portion of Minnesota. Congress had already ratified the peace treaty with Britain in 1784 and appointed Thomas Jefferson chairman of a committee to dispose of the land.

It is vital to understand that from the beginning the intention of the founding fathers was to sell this land for revenue and to pay the enormous debts created by the war. Consider the situation: a sparse population strung out along the eastern seaboard; limitless land stretching westward, for the most part unmapped, unexplored, unknown. More land than a man could ever imagine being used. Moreover, it was the one asset of the new nation that could be converted into cash.

To dispose of it the Land Ordinance of 1785 set up a basic pattern of 640-acre sections in square grids. One section in each township was earmarked for education and another portion set aside for veterans' claims. The rest was to be sold at a price not less than one dollar an acre. This basic 640-acre grid is still the standard of the Federal land system today.

Men like Jefferson understood that without westward expansion, with Spain, France and England holding territories around it, America could not become a great nation. Jefferson was aided by the continuing wars in Europe, which diluted the interest of the warring nations in America. So acquisitions followed fairly rapidly. The Louisiana Purchase in 1803 added more than 500,000,000 acres; the Oregon Compromise in 1846 another 180,000,000 acres; the huge area of the Southwest, taken by conquest from Mexico in 1848, brought 334,000,000 acres; Texas was annexed in 1845, and in 1850 a purchase from Texas added 78,000,000 acres, including parts of Colorado, Kansas, Oklahoma and New Mexico; the Gadsden Purchase in 1853 brought nearly 19,000,000 acres; and, finally, there was the purchase of Alaska from Russia in 1867, with 365,000,000 acres.

The average cost was about five cents an acre. Alaska, a real bargain, came for two cents an acre. Total: about 2,000,000,000 acres.

From the close of the Revolution until the passing of the Homestead Act in 1862 about 65,000,000 acres were used to pay Federal debts to individuals. Claimants were given scrip valid for a designated quantity of land. An additional 225,000,000 acres were transferred back to the states to support schools and institutions and to be used for related purposes. The Federal government still owes 15 states about 1,000,000 acres in unpaid debts.

The great gravy train operated with little friction for a long time—as long as there was more land than claimants. The Homestead Act allowed settlers to stake out claims and establish permanent ownership by occupying and improving their land. Under this act a million and a half people settled 248,000,000 acres of the West. Though never repealed, the Homestead Act has been more or less superseded by other laws, such as the Taylor Grazing Act of 1934, which withdrew land from settlement, so that by 1969 only 69 new land titles were granted, most of them in Alaska.

The Homestead Act permitted 160 acres to the claim. In an arid climate, like that in much of the West, this was not enough to support either livestock or farming. It was followed in 1877 by the Desert Land Act, which permitted the purchase of 640 acres (later reduced to 320 acres) if they were irrigated. Recognizing that small farmers could not afford the cost of irrigation by themselves, Congress then passed the Carey Act, which transferred 1,000,000 acres to each state in the arid West that would encourage and help farmers in improving their land, presumably by irrigation.

In 1902 the National Reclamation Act was passed, making the Federal government responsible for reclaiming public lands, or even privately owned lands, if the costs were borne by the landowners. A flood of land acts continued to pour out of Congress—one estimate is 5,000.

The Stock Raising Homestead Act, passed in 1916, permitted homesteads as large as 640 acres if the land were used for grazing. To stimulate railroad building, the government donated 94,000,000 acres to the railroad builders in the form of rights-of-way, plus alternating sections of land for 20 miles on each side of the railroad line. These tracts were sold by the railroads to raise money for building their lines. Another 37,000,000 acres were given the states to be sold for financing railroad construction. It goes without saying that the rights of the Indians were never considered at all.

The checkerboard grid pattern created an interesting monopolization of land that was very characteristic of the old wild West. It led to interesting conflicts, the basis of much fiction, legend and real history.

Purchasers, or homesteaders, would select key portions of land containing a water source or water rights. As a result no one would buy or settle the adjoining territory, because it was useless without water. The owner of the water rights could thus enjoy free use of all the land around him, since he controlled the water for all of it.

This pattern helped establish some of the great ranching empires of the West, founded on the free use of public land. The conditioning of usage was so strong that many ranchers came to believe the land was theirs, and actually fenced it. By 1964 about half of the Bureau of Land Management's domain in eleven western states was fenced. Not only was it fenced, but the bureau actually paid two-thirds of the cost—so that the public paid to keep itself out of public lands!

The next problem of the public domain was concerned with the rights and privileges of miners. Prospectors filed claims on public land without hindrance, except occasionally from the Indians—whereupon the Cavalry was called in to arbitrate the disagreement, with the usual disastrous results for the Indians.

Miners drew up their own body of regulations, which, in essence, provided that the discoverer of a mine or the parties to whom he disposed of it were entitled to the benefits of the development.

The first mining laws, which came in 1866 and 1872, followed the general pattern, holding that a claim belonged to the discoverer, but title hinged on development. A patent cost $2.50 an acre for a placer claim and $5.00 an acre for a lode claim.

No one had yet coined the phrase "multiple use," and these laws were strong for single use. They said in effect that if valuable minerals were discovered on public land, the exploitation of the mineral rights superseded all other rights. The principle was nailed down in a historic case in 1894.

A homesteader named Martin Womble had filed on 160 acres. Before his claim could be finally proved, a miner named Walter Castle located a mining claim on the same property. Womble filed objection, and the dispute, according to the Mining Law of 1872, boiled down to the question of value of the minerals in the claim. Womble argued that the minerals on the disputed land were not valuable enough to upset his homestead claim as required by the law.

The decision was made by Secretary of the Interior Hoke Smith. Known as the "prudent man" rule, it set the standard: "A mineral discovery, sufficient to warrant the location of a mining claim, may be regarded as proven, where mineral is found, and the evidence shows that a person of ordinary prudence would be justified in the further expenditure of his labor and means, with a reasonable prospect of success in developing a valuable mine."

The case went all the way to the Supreme Court, was upheld, and the law of 1872 is still the governing principle in filing mineral claims, antiquated as it may be.

The results were sometimes ludicrous. When minerals were discovered in the Tucson area of Arizona, prospectors swarmed into what had become a residential area and filed claims in homeowners' yards—even began digging up their lawns. To the amazement of the homeowners, the law protected the miner, regardless of the amount of damage he was doing to private property. It took a special act of Congress to exclude Tucson from the general rule of mineral entry.

Early miners, of course, were most interested in gold and silver; later on, the problem became more complicated. How did oil, gas or coal figure in mining claims?

Upon the recommendation of the Department of the Interior, the lands believed to contain oil, gas, coal or phosphates were withdrawn from open claim under the mining law. The Mineral Leasing Act of 1920 established a system of leases and payment of royalties for the exploitation of these resources and added sodium and potassium—later sulphur, asphalt and bitumens.

The Taylor Grazing Act of 1934 set up grazing districts in the West under Federal administration for 142,000,000 acres, "pending their final disposition." The phrase has been construed to mean that Congress ultimately intended to dispose of all the public domain. Meantime Federal grazing fees were set so low as to constitute an actual subsidy to stockmen, in much the manner that the Mining Law of 1872 was an outright subsidy to prospectors.

As the once limitless West was settled and shrank, a few farsighted men began to advance the outrageous notion that the cornucopia was not bottomless and that perhaps it was not too soon to begin thinking of saving parts of the public domain, either for the future or just to protect it from needless destruction. Conservation, or protectionism, was born with the first national park—Yellowstone, established in 1872. The intent clearly was to protect and preserve the natural landscape from exploitation.

The following year the Timber Culture Act was passed, under

which a tract of 160 acres could be granted to a claimant who would plant at least 40 acres in trees and maintain it for 10 years. This law failed and was repealed in 1891, but the repeal carried a rider that provided for the establishment of national forests on public lands.

Forest reserves were fairly successful. About 50,000,000 acres were accumulated by a succession of Presidents from Harrison to McKinley. Theodore Roosevelt added another 150,000,000 acres, and the Weeks Law, passed in 1911, legalized the creation of forest reserves in the eastern states, where there was no public domain.

Roosevelt was also responsible for founding the National Wildlife Refuge system in 1902, with the Pelican Island Refuge in Florida. The system now covers 27,000,000 acres, most of it in Alaska and a good deal of it under constant pressure from special interests who want hunting rights, timber rights or oil rights. As an example, the Kenai National Moose Range in Alaska contains timber, oil, gas, coal, sand and gravel, and agricultural land, as well as moose, bear, Dall sheep and mountain goats. The pressure is on to develop the area, and a number of leases for oil and gas exploration have been filed despite their illegality, leading to some tangled litigation and the inevitable remark by an Alaskan politician, "The Kenai is for people, not for goddam moose!"

A rough accounting of the present public-domain situation is about as follows:

NUMBER OF ACRES	
61,000,000	granted to veterans
94,000,000	to railroads
287,000,000	to homesteaders
330,000,000	to the states
34,000,000	to private claims
302,000,000	cash sales
14,000,000	sold under Timber and Stone Law
11,000,000	sold or granted under Timber Culture Law
10,000,000	sold under Desert Land Law
135,000,000	classified for retention and multiple-use management, of which 110,000,000 acres are in 11 western states and 25,000,000 in Alaska

The remaining acreage awaits "final disposition." Cries of fraud have been relatively few, but over the years a difference of opinion as to what should be done with the public lands, and a controversy, have been building up. Since so much of it is in the western states (Nevada is 86 percent public domain), the states would very much like to get hold of these acres and put them on the tax rolls. Stockmen, of course, want more land for grazing, and so far have been able to keep Federal grazing fees below local rates. The mining industry would like to retain or even enlarge the permissive doctrines of the 1872 law, giving prospectors access to minerals anywhere on public lands. The timber industry made its bid for increased cutting in the national forests and was, temporarily at least, turned back, though it has not given up, and its arguments are being given more than sympathy by President Nixon. And, finally, the conservationists would like to see more of the public domain brought into the wilderness or parks systems.

Two kinds of conflict have developed. Private enterprise had its grievances. The mining industry complained that often it could not carry on large-scale operations, because even when its claims were legalized, it was hampered by acreage restrictions. Ranchers using public land refused to make improvements or even protect it adequately, and a massive deterioration of grassland resulted. Towns anxious to expand found Federal land blocking the way and resented the loss of tax revenue represented by the land denied them.

On the other hand, the huge public domain attracted campers, hunters, and fishermen in steadily increasing numbers. Not all of them were wilderness-oriented and were annoyed by the lack of conveniences and sanitary provisions. The result too often was litter, vandalism and destruction.

If the general public was unaware of all this, Congress was not, and pressures to do something about the Bureau of Land Management grew steadily. The bureau was criticized for duplicating the functions of other agencies, particularly the Forest Service, and it was suggested that it be merged or combined with other agencies concerned with land management. And the question was asked, over and over: What is to be the final disposition of the public domain?

In his environmental message of February 10, 1970, President Nixon had recommended the establishment of a Federal Property Review Board, "to recommend to me what properties should be converted or sold." The proposal was criticized by some congressmen as a means by which Federal land could be as easily lost as protected. The American Forestry Association and allied groups had formulated a series of principles to apply to the management of public lands, each of which had to be gauged by a single criterion: Is it in the public interest? As to disposition, the conclusion was, "Disposal of federal lands should be permitted only when demonstrated public needs indicate a higher public service will be achieved." Such a principle might make it a little difficult to demonstrate that the public service would be benefited by allowing a few people to profit at the expense of the rest of the public.

And along those same lines, there was general agreement that the Mining Law of 1872, permitting the filing of a claim anywhere on public land, should be repealed and replaced with a mineral-leasing system. Other laws crying for repeal were the antiquated Homestead Act, the Desert Land Act, and the Small Tract Act, under which bits and pieces of Federal land have been sold, often cutting up public domain into a hodgepodge jigsaw puzzle.

Other conservation objectives called for making BLM lands candidates for classification under the National Wilderness Act. The Wilderness Act of 1964 did not make any provision for including the public domain under the National Wilderness Preservation System, but this does not mean that the Bureau of Land Management could not so classify its lands. However, it had not done so, nor paid much attention to an interpretation of the Multiple Use Act, which would give wildlife and recreation as much weight as commercial uses.

Other objectives considered important by conservationists dealt with halting any erosion of the public rights on public domain. For example, it was considered necessary to affirm the principle that users of public land do not acquire or retain any rights in the land they are using, or any equity or reimbursement when their permit to use expires; above all, that funds and manpower be supplied to protect public lands from vandalism, to halt exploitation, and to enforce sustained growth of forests.

The problems of the public domain have particular poignancy in Alaska. It is our only frontier state, and it contains 280,000,000 acres of public land. Thus Alaska still has a chance—a very slim one, it seems—to avoid the costly blunders committed by the other states. The Statehood Act provides 104,000,000 acres for the state to use for development. So far about 18,000,000 acres have been chosen, and these have been picked primarily for exploitation of oil or mineral resources, with the same lack of concern for damage to the environment or for future needs. The probability is overwhelming that political pressures in Alaska will force the same pattern of use and destroy for short-term profit that has gone so far in the other states.

The various pressures had culminated by 1964 with the creation in Washington of the Public Land Review Commission, empowered to examine all existing laws and regulations, review Federal policies and agencies dealing with public lands, and recommend changes.

Representative Wayne N. Aspinall of Colorado was elected chairman, and Milton A. Pearl, a lawyer and real-estate specialist from the House Interior Committee, was named director. Six members were appointed by the Speaker of the House, six by the president of the Senate, and six by the President of the United States. In addition, there was an advisory council of 33 members, plus representatives of the governors of the 50 states.

With a budget of $7,000,000 for a projected five-year span of operation, the commission heard 900 witnesses and took 5,000 pages of testimony. The report was due on December 31, 1968, but was not actually made public until June 24, 1970.

In those years when the commission deliberated over "the chaotic jungle of land laws" dating from 1792, there was surprisingly little speculation among conservationists as to its ultimate recommendations, in spite of their misgivings about Chairman Aspinall. The Representative from Colorado has been called a symbol of "an aging, unresponsive House establishment." As chairman of the House Interior Committee he has been in a position to block important legislation on conservation matters. His critics, said Marjorie Hunter in *The New York Times,* "regard him as a protectionist for western mining, timber and grazing interests, tending to favor private commercial exploitation of public lands and resources."

It should have come as no great surprise, therefore, that the commission's recommendations, finally revealed in June, stood four-square for such changes in the management of the public lands as would give private interests a bigger cut of the pie.

"The report," said Gladwin Hill in *The New York Times,* "while repeatedly stressing judicious 'multiple use' of public lands with solicitude for environmental values, hewed closely to policies advocated by the timber, mining and grazing industries, which conservationists have denounced as overly exploitative."

It paid not even lip service to the conservation of exhaustible resources such as oil, coal or minerals, leaving this matter to "the normal operations of the market place."

The implications of that phrase are a little incredible, but no more so than a following remark by Milton Pearl, the director, at a news conference. "The Commission," he said, "saw no reason for superimposing the views of government executives on the decisions of business executives."

What price guardians of the public weal!

The constant concern of the commission, said Chairman Aspinall, was to balance commercial uses of the public lands with environmental and recreational objectives. This, he said, was constructive, because "nature is one of the worst offenders in regard to maintaining environment."

Consider, then, our good fortune in having Aspinall to protect us from the ravages of nature. His place in history is secure.

After the symbolic pat on the head for "multiple use" the commission turned its back on it and recommended that public lands highly productive of timber be classified for commercial timber production *as the dominant use.*

And, having played Santa to the timber industry, and not wishing to slight the miners, the commission said: "Mineral exploration and development should have a preference over some or all other uses in much of our public lands. We recognize that [it] will in most cases have an impact on the environment or be incompatible with some other uses."

To nail down this gift to the mining industry, the commission balked at repealing the Mining Law of 1872, a law that everyone but the miners has wanted repealed as a "scandalous device" for the illegitimate acquisition of public lands. The most the commission would consider was some revision of the law.

The implications of the study are grim. If the congressional outlook is no more progressive than this, there is scant hope of saving our precious, limited land resource. A study that took five years and cost more than $7,000,000 has done little more than reassert the doctrines that have resulted in so much destruction of the American earth.

Meantime other events continued to stagger through history. In May 1963 Congress passed a law that recognized the desirability "that all American people of present and future generations be assured adequate outdoor recreation resources." The bill directed the Secretary of the Interior, then Stewart Udall, to formulate "a comprehensive nationwide outdoor recreation plan," and have it ready in five years.

Five years later Interior asked for, and received, an extension of time. But whatever plan it was formulating is now entombed in the archives, because the Nixon Administration took over in 1969, and Walter Hickel became Secretary of the Interior, with definite ideas of his own.

Secretary Hickel made it known that his plan was to bring the parks to the people through a massive program of grants to states and cities. This was to be done in two steps.

In a five-year urban recreation program, cities of 250,000 or more people would get Federal grants to finance recreation projects on "an imaginative and massive scale"—up to 90 percent of the cost. The Federal contribution was put at $5,300,000,000, on top of the $200,000,000 already authorized for park acquisition. It should be noted that this $200,000,000 was authorized, but none of it was spent, because authorizing and actually getting the money are two very different things on Capitol Hill.

The second step was a $1,000,000,000 program for the National Park Service to acquire, develop, and operate urban areas of outstanding quality.

Thus the total cost of the Hickel program came to $6,300,000,000, and it promptly mired down in the Budget Bureau, where it remained. A thick book of about 1,000 pages, titled, "The Recreation Imperative," was on the press but was hurriedly taken off when the impasse in the Budget Bureau became known. Nevertheless Secretary Hickel spoke publicly of his plan, and in the latter part of 1969 the Bureau of Outdoor Recreation published a booklet with text and pictures on "the recreation imperative." It contained the presentation made by Secretary Hickel to President Nixon. And, in one of the last times that the President may have listened to the Secretary, Nixon incorporated part of the Hickel plan in his environmental speech of February 10, 1970. That speech proposed legislation permitting the Department of the Interior to transfer surplus public land to states and cities for parks.

By this time the idea was not exactly new to Congress. Senator Henry M. Jackson of Washington had introduced such a bill in 1969; it had passed the Senate in June and was pending in the House.

The need for expansion in the National Park System had grown painfully obvious with the boom in camping. The problem is that most of the millions who have lately discovered the lure of the outdoors are not backpack campers but automobile campers. They come to the parks in trailers and truck campers, and they require parking space, shopping facilities, electricity, water and policing.

The result has been to turn the parks into cities rather than camps —worse, into slums, where congestion, dirt, noise and crowding match the worst of the city. Although the National Park System has grown from 16 parks and 22 national monuments in 1916, when it was established by Congress, to 35 parks and 237 areas, such as national seashores, recreation areas, battlefields and memorials, it has still fallen behind the demand.

George B. Hartzog, Jr., director of the National Park Service, notes that when the park system was established an average wage was $708 a year and few workingmen received paid vacations. Most people didn't travel, nor could they afford the high-priced equipment now clogging the roads.

By 1969 the average income was $8,017 a year, with the five-day week standard, and with nearly everybody, office or factory worker, getting a two-week vacation after five years—a very large number getting three weeks after 15 years. Many people get four weeks vacation after 15 or more years.

Visits to the national parks and monuments rose right along with this increased leisure time. Superhighways and faster cars enabled tourists to reach almost any park in the country in a few days. In 1958 there were 65,700,000 visits to the country's national parks. In 1969 this had grown to 197,000,000 visitor days (40,000,000 to the parks and 157,000,000 to national forests). From 1960 to 1968 the population increased by 9.4 percent, but visits to the parks increased by 90 percent. They are still increasing at the rate of 7 percent a year.

The resultant crush not only made camping in the national parks and forests unpleasant but brought some real problems. Traffic jams on the roads have caused backups lasting five hours. Highway 441, which crosses the Great Smoky Mountain National Park, has had 20 miles of stalled traffic on a summer weekend.

Campsites are being pounded into desert by tires and feet. The smoke of cooking fires has created a smog condition like an industrial town's. Yosemite has already had its inversions, when smog was so bad that climbers at Glacier Point, 7,200 feet high, could not see the valley below them.

Sewage facilities are overtaxed, and at Yosemite raw sewage has been poured into the Merced River. Vandalism grows—trees have been hacked and damaged, signs knocked down, benches destroyed, doors of service buildings ripped from their hinges, campfires gone out of control.

Even crime has become a problem, rising 67.6 percent in the parks in 1968, as compared with 16 percent for the country as a whole. The Yosemite Rangers find themselves serving as police instead of naturalists, and they resent it. While vandalism and pilfering are common, other crimes range from felony and assault through sex offenses and murder. Juvenile delinquency is a major problem, with Rangers at Yosemite being forced to train as riot squads to break up mobs of destructive teen-agers. Liquor is another problem, and the Rangers spend a lot of time confiscating beer and hard liquor, which by Labor Day fills an empty room to the ceiling in park headquarters.

The newest form of vandalism utilizes a spray can of paint, rather than a knife, to apply names or graffiti to rocks, walls and trees. The Rangers have been able to catch few of these artists.

Yosemite is our most popular park and the most crowded, with 2,500,000 visitors in 1970. While the park itself is large, covering 1,200 square miles, the great influx is into the seven-square-mile Yosemite Valley, where some of the country's most breathtaking scenery is concentrated.

As early as 1963 the Curry Company, which operates the hotels, campgrounds and other concessions in Yosemite, was advising reservations a year ahead. The prospect today is closer to five years ahead, which gets to be a little ridiculous.

From the concessionaire's standpoint, the increasing crowds have merely created a kind of mixed bonanza. The small valley is now crowded with nine grocery stores, seven gas stations, three swimming pools, a coin-operated laundry, a barber shop, a stable for 300 riding animals, several bars, and at least 4,500 hotel rooms, plus cabins, tents and several huge campgrounds. But match a total of 5,148 sleeping accommodations, or "pillows," as they are called, against 70,000 visitors on a holiday weekend, and you have chaos.

"The whole valley will be black-topped soon," said John Hansen, assistant postmaster, gloomily in 1966. "People come here to-day to be entertained—they expect popcorn stands and Ferris wheels."

As before unhappily observed, not everyone comes for the peace and quiet of the High Sierras. Those who still do are driven away by the incessant throb of engines, the clamor of transistor radios, and the churning hubbub of large masses of people in constant motion.

"Those who come now don't mind the crowds," glumly observed Chief Naturalist Bryan Harry, "in fact they like them. They are sightseers, and they come for the action. They don't come for what Yosemite really has to offer."

He pinpointed the problem. "Yosemite will still be the same size fifty years from now. We can't make it any bigger or build another one. The population is not only growing, it is becoming more affluent and more mobile and this land will become even more precious as other wild places continue to shrink. We have to find a way to cope with this problem—and we are open to suggestions."

The suggestions made are considered generally unpalatable to a Park Service anxious to accommodate everyone. The "No camping space available" sign is out more and more frequently. And the suggestion made for the cities is now heard more loudly for the parks—keep the cars out.

A start was made in 1967, when the Curry Company inaugurated a shuttle-bus service to transport tourists from one point to another for 25 cents, or all around the valley for 50 cents. In September 1969, cars were barred from the Mariposa Grove of sequoias and replaced by sight-seeing buses. And in July 1970, cars were banned from the eastern end of the valley in a preliminary step toward a total ban for all of Yosemite Valley. However, the campsites are still open to cars and trailers, although the number has been arbitrarily reduced by half.

But if auto camping is eventually barred, where do the campers go? There is public or private land outside the parks that can be developed as campgrounds. Private landowners generally welcome the opportunity for income to be derived from accommodating campers and would make the investment in facilities required.

From the park gateway, entry can be on foot, horse or shuttle bus. This would, at one stroke, end the traffic jams, eliminate the campground ghettos, and help protect the fragile ecology of such areas as Yosemite Valley. Even the building of monorails into the parks has been suggested. If this seems like an alien intrusion into wilderness, consider that a monorail would actually disturb the ecology less than roads, be quieter, create no pollution and interfere far less with wildlife.

Unless more parks are created, rationing indeed may become a reality. What is most important to decide is the function that parks should serve. Those who want popcorn stands and Ferris wheels can be accommodated in recreation areas closer to urban centers, with the wilderness left for those who can appreciate it and use it wisely. National parks, said the report of the Secretary of the Interior's advisory board on wildlife management, should be a "vignette of primitive America," with "the biotic associations within each park . . . maintained, or where necessary recreated, as nearly as possible in the condition that prevailed when the area was first visited by the white man."

There are proponents of progress who take a dim view of all this clinging to the past and who cherish no sentimental attachment to an America in its pre-white-man condition. They fail to understand the basic meaning of the conservation movement. The desire to save some of primitive America is not mere sentiment, although it is only honest to admit that sentiment plays a part. But the motives are more serious than this. Conservationists believe that the preservation of wild America—or as much of it as can be preserved—is essential to our physical and mental health, that the need for getting away from the steel and concrete of civilization is an important need, even if it is distorted by the auto campers. Furthermore, an undamaged area of wilderness is an important ecological laboratory that can reveal facts about our relationship to the rest of our environment that may be essential to our survival—assuming we learn them before we destroy the planet. As a Conservation Foundation report, "Man and Nature in the National Parks," observes, "The fact that few people understand the scientific detail of the ecology makes no difference to the assumption that a landscape in ecological repose is generally one that gives pleasure." Unless an informed biological policy is followed in managing our parks they will deteriorate as they have, to the point where they no longer are what they were supposed to be. Finally, and not least, we owe future generations at least a token look at the world we came into and have so drastically altered.

Meantime park officials, not overly optimistic, appear to believe that even if more land is made available for parks, their use will have to be rationed eventually. A study is now under way to try to determine the saturation point at which a park environment starts to deteriorate. When that point is reached, the alternatives are equally unpleasant—the destruction of the park or limited access for the public.

In the last decade Congress has added a number of new areas to be administered by the Park Service, including seven national seashores. The first attempt to save threatened areas of the shoreline came with a survey made in 1935. This survey recommended that 12 areas, with a total shoreline of 439 miles, be preserved in the national park system.

Considering the thousands upon thousands of miles of shoreline that form the perimeter of the continent, a goal of 439 miles is microscopic; yet only one area was actually acquired by 1937—Cape Hatteras. It provided 70 miles of ocean front. It took another 20 years before another survey was made of seashore recreation areas along the Atlantic and Gulf coasts, as a result of which 16 areas were designated for acquisition. Three of these—Cape Cod, Fire Island, and Padre Island (Texas)—were actually acquired over the next 15 years. Meantime the remaining shoreline was being bought up at top speed for industrial and private development.

Some ecologists feel that offshore islands should be acquired as national seashores, affording the best opportunity to preserve the unique edge-of-the-sea ecology. Many of these islands are owned privately and represent difficult acquisitions for the Park Service. In recent years, however, the ownership of the islands has passed to the heirs of the original owners. In some cases they might not have the same interest in the property, affording the Park Service a better opportunity to buy and preserve the islands as wildlife or recreation preserves.

The seven operating national seashores are plagued with the same problems of overcrowding and litter as the other parks. Assateague Island in Maryland, where tourists come to see the wild ponies, "gets so jammed in summer," said Park Ranger Leonard McKenzie, "that we can't cope with all the people."

Assateague prepared for 800 campers on the July 4 weekend of 1969, but 1,400 arrived as early as Memorial Day. "We're going to have to reduce the number of our camping sites," said McKenzie. "We hate to do it, but we can't preserve the environment and provide a quality camping experience with such large crowds."

At Cape Cod, a 55-mile preserve, there were 4,000,000 visits in 1969, and the total has been growing at 10 to 15 percent a year. "One of our main problems," said civil engineer James H. Bowman of the park staff, "is the impact of people on the dunes—even a footprint can lead to a deep ravine in a dune if the wind comes up and blows the sand away."

The story is the same clear across the continent to Point Reyes in California. Point Reyes is particularly unfortunate, being a crazy-quilt patchwork of public and private tracts, with subdivisions going up, bulldozers churning out new roads and "Keep Out" signs cropping up everywhere to confuse the camper who thinks he is in a national park.

The confusion stems from the fact that not enough land was purchased at Point Reyes. It is "too fragmented," says Undersecretary of the Interior Russell Train, "and insufficient to be regarded as efficiently administrable."

This is something of an understatement. The Park Service estimated $14,000,000 to acquire the land. Inflated land values sent prices soaring, and about $38,000,000 more is required to complete the purchases, which means that the present parkland is chopped up and interspersed by all kinds of private developments.

Originally this did not seem to be a problem. President Johnson had signed the Land and Water Conservation Fund Act in 1968 which authorized $200,000,000 a year for five years to acquire scenic areas. In the last months of his administration Mr. Johnson cut the amount to $154,000,000. When the new administration came in President Nixon cut another $30,000,000 out of the fund. Of the remaining $124,000,000 only $14,000,000 was left for the National Park Service, and there were other priorities that simply washed out Point Reyes.

The Sierra Club, incidentally, took public notice of the cuts by an angry comment from its vice-president, Dr. Edgar Wayburn. Said Dr. Wayburn at a news conference in October 1969, "Since President Nixon took office last January he has not recommended to

Congress a single addition to the national parks system or the wilderness system."

To get around its poverty the Park Service is flirting with an idea they think may solve their problem, but which involves some financial juggling. To deal with the $38,000,000 shortage at Point Reyes, they propose to ask Congress for $28,000,000, with which they will buy all the land to complete the park in a deal that calls for leasing back to farmers 16,500 acres and selling outright 9,200 acres to urban developers to build homes, stores and gas stations as they see fit, thus raising enough money to cover their debit. If the plan works at Point Reyes, the Park Service will consider extending it to Cape Cod and perhaps other parks, such as Yosemite, Glacier or Sequoia.

Is it ethical? The government would be in the position of asserting that it was protecting a land area from commercial development by buying it (through condemnation proceedings if necessary) from the original owner, then selling it to someone else and encouraging him to do the very thing it had just prevented the original owner from doing. The new purchaser would actually be assured of a profit whether he used the land or resold it, because it would now be part of a national seashore or park.

Apart from the questionable ethics, conservationists view the plan as a step backward in park development. It would create land holdings inside a national park or seashore, a situation the Park Service has been trying to eliminate for years. As Director Hartzog has said, such inholdings are like worms in an apple, not taking up much room but spoiling the fruit. Buying up such pieces of private land inside a park becomes more expensive such year. At Everglades National Park there are some 70,000 acres of private holdings that would now cost $20,000,000 to acquire.

Rising prices and declining appropriations have placed the entire matter of new parks in jeopardy. Cape Cod National Seashore was established in 1961, but only a small part of it, divided among four separate areas, is open to the public. On the record, the Park Service seems to have had its greatest success in establishing new parks where the land is already owned by the public. Which brings us full circle back to the Bureau of Land Management and the 755,000,000 acres in the public domain. Why can't parks be given at least equal priority with the demands of the mining and timber industries?

CHAPTER

11

Protection or Multiple Use?

IT is time we took a closer look at the principle of multiple use. As applied to public lands, there is disagreement and acrimony between conservationists and those industries that feel entitled to use public lands for commercial purposes—understandably. There is even a certain level of disagreement among conservationists themselves. One end of this spectrum is labeled "protectionist" or "preservationist" and is accused of being made up of wilderness nuts who wish to "lock up" vast areas and prevent their "development."

Differing from them are the recreationists, who believe that under the mounting pressure for open space, public lands should be developed or "improved," with roads and convenience facilities, such as picnic areas, parking lots, swimming pools, comfort stations, food concessions and so on, so they can be fully utilized by the motoring public.

Recreation, certainly, is one of the factors of multiple use, which also includes lumbering, mining, grazing, hunting, fishing, camping, hiking and mountain climbing, swimming or just loafing. It might also permit the building of summer camps, hotels, ski lifts, toboggan slides, nature trails or amusement parks. And multiple use, as we have seen, has been raised to the status of a creed by many who use it as both a principle and a weapon. However, like any orthodoxy, multiple use runs into strange and unforeseen problems.

On May 6, 1969, the U.S. Forest Service in Colorado offered for public sale the timber it had marked for cutting on 357 acres of public land at East Meadow Creek, on the western slope of the Gore Range.

Bids on 4,300,000 board feet of timber, invited by newspaper advertisements, were legal and within the responsibilities of the Forest Service under the Multiple Use-Sustained Yield Act of 1960. Notwithstanding, on April 4 a citizens' lawsuit to stop the sale was filed in the U.S. District Court at Denver. The suit named as defendants Secretary of Agriculture Clifford Hardin, Chief U.S. Forester Edward P. Cliff, Denver Regional Forester David Nordwall, White River Supervisor James O. Folkestad, and the Kaibab Industries, an Arizona lumbering company that took up the bid.

The suit was a first of its kind protesting a Forest Service decision. It was filed by 12 citizens of the nearby resort town of Vail, the Town of Vail, the Sierra Club, *Colorado* Magazine, the Colorado Open Space Coordinating Council (representing 27 groups), and William B. Mounsey, wilderness consultant to COSCC.

This lawsuit was not an impulsive act. It was the culmination of ten months of frustrating activity and agitation, during which the citizens directed questions at District Ranger Donald E. Price and maintained a feverish correspondence with the Forest Service, asking for clear answers as to the advisability of the sale. The answers they got were soothing but unsatisfactory, with the result that they formed an ad hoc Eagles Nest Wilderness Committee and filed suit to stop the sale. But on what grounds, if the action was legal and within the admitted jurisdiction of the Forest Service? Their motivation throws an interesting light on the application of the multiple-use principle to public lands.

The area around East Meadow Creek covers about 2,400 acres surrounding the stream that flows out of the Gore Range–Eagles Nest primitive area of the White River National Forest. The town of Vail is eight miles to the south, a noted ski center and forest area, where lumbering and other elements of multiple use receive full treatment.

East Meadow Creek itself is not in the primitive area of the White River National Forest but just adjacent to it and in a state of de facto wilderness. Gore Range, a primitive area, was scheduled for inclusion in the Wilderness System by 1970 or 1971, and, as is usual in such cases, the contiguous area of East Meadow Creek, with wilderness potential, would also be considered. In the past, new wilderness areas established by Congress have included about a quarter of their acreage from such surrounding primitive areas.

In the normal course of events, therefore, East Meadow Creek would have been destined for inclusion in the Wilderness System. *But not if it were logged and roads cut through it.* And this was what had fired the protest.

Why, the citizens of Vail wanted to know, was the Forest Service rushing to have it logged when such action would ruin it for wilderness inclusion? It was just a little suspicious, they felt, that Forest Service timber sales seemed always to be weighted toward those areas that were similar candidates for wilderness reclassification. The suspicion had created what local people called "a credibility gap" between the Forest Service and themselves.

It wasn't that the decision to sell the timber was wrong, said Roger P. Hansen, director of COSCC. The process by which the Forest Service arrived at the decision to sell was illegal. And that was a different thing altogether.

The law says that before making a decision the Forest Service must conduct studies to balance all the possible uses of the land and determine the relative weight of each use. The Forest Service must also make wilderness studies on lands that are likely to be added to the Wilderness System. These studies, said the plaintiffs, were not made. If all studies required had been properly done, they would have shown that lumbering operations in this area were uneconomical anyway. And so the fate of East Meadow Creek is stalled in the interminable traffic of the law courts.

In these situations the Forest Service is caught in a cross fire, with the lumber companies demanding action on timber sales and often utilizing political pressure. An unhappy Forest Service man put his finger accurately on his problem when he said, "Everyone believes in multiple use as long as it's *his* use!"

Which sums it up well enough, but leaves unanswered a basic question: for whom is the Forest Service working—the public or private interests?

In Idaho there is a primitive area at the headwaters of the Salmon River named White Clouds. It is a region of tremendous snow-capped peaks, deep valleys carved by glacial action, and many clear, cold lakes.

The highest mountain in this area is Castle Peak, and at the foot of Castle Peak is the base camp and mining claim of the American Smelting and Refinery Company. It is perfectly legal for American Smelting or anyone else to prospect for metal on public land. The company geologists have located a deposit of molybdenum and American Smelting has announced its intention of taking out the metal by open-pit mining, one of the most destructive of all activities to the landscape.

There are no roads in this primitive area, and teams of 30 to 40 specialists have been airlifted in and samples taken out by helicopter. The company says it has spent a million dollars in geologic exploration and drilling. As a first step for its proposed operations, American Smelting applied to the Forest Service for a permit to build an eight-mile access road to reach its 740-acre claim. Under the law the Forest Service has no choice but to grant the permit. It has held three public hearings, one in Boise, one in Idaho Falls and one in Challis.

Governor Don Samuelson of Idaho professed himself delighted with the proposed mining venture. "The raw materials are here and should be exploited to the fullest possible extent." Since Idaho is 64 percent Federal land, and the remaining 36 percent of private land pays the real-estate taxes, he welcomed all such ventures as this to broaden the tax base.

Expressing the sentiment exactly opposite, Ernest E. Day, chairman of the Idaho Parks Board, resigned with a bitter thrust at the governor. "I don't see any sense of being part of a team which doesn't have enough regard for our resources to better differentiate between them."

American Smelting says the market for molybdenum, an alloy used for toughening steel, is growing at the rate of 7 percent a year. The opposition, which includes the Wilderness Society, the Idaho Alpine Club, the Greater Sawtooth Preservation Council, the Idaho Wildlife Federation and a dozen others, say that the grade of ore at Castle Peak is low, that the metal is not in short supply, and that the damage done by open-pit mining will far outweigh the value of the metal taken out.

American Smelting admits that molybdenum is not in short supply but maintains it is needed to trade for other metals in world markets. Besides, says the company, it has a perfectly legal right to come in and work its claim, and there are laws protecting free enterprise.

The opposition grants that the law permits mining companies to work claims on public land, but they bitterly question the legitimacy of such operations on the grounds that the public should have the right to decide what happens on public land.

Lumber companies at least must *buy* the trees they cut. Mining companies don't even have to do that—they may just take the minerals. They can come in, tear up the land, pollute water and air, and then move on, leaving destruction behind. Said Idaho's Congressman Orval J. Hansen, "It is clear that the 1872 Federal mining law does not adequately protect the public interest. The law should be amended to provide that all values, including scenic, recreational, fish and wildlife, as well as economic benefits, be considered before mining operations are authorized on public lands. Such a change in the law is long overdue."

Governor Samuelson denied that mining would create stream or lake pollution. "We have laws to protect that," he said. But it is acknowledged that American Smelting will dump its tailings into the nearby lake; the company has promised to "restore" the lake once the work is concluded. On the other hand, a company official has said that the deposit contains many millions of tons of ore and the operation would have a long life, so just when the pollution is to stop and the restoration to begin remains problematical. Since the yield is less than half of 1 percent, more than 99 percent of the rock dug out would simply be dumped as waste, and it is not difficult to visualize the alteration in the landscape resulting from the dumping of 20,000 tons of rock a day.

"The sheer dimensions of a molybdenum operation," said the Boise, Idaho, *Statesman*, "recovering less than one-half of one per cent of millions of tons of ore, creates an immense obstacle to conservation."

There is danger of pollution, not only of the lake and the Little Boulder Creek, but of the Salmon River, one of the last of the western rivers where steelhead and salmon still come to spawn. Furthermore, American Smelting's request for a road permit speaks of "development and further exploration," which could mean new claims and further inroads.

Local residents favor the operation by a margin of 25 to one.

American Smelting has said it will spend about $40,000,000 in capital expenditures, employ about 350 men directly and another 500 indirectly. These are persuasive arguments to the governor and the Chamber of Commerce.

Among the dissenting minority is rancher Lawrence Kratz. If there is no scarcity of molybdenum as everyone grants, he says, why not leave it where it is? It will still be there 25 years from now, and if by then a real scarcity has developed, it can still be mined. "In the meantime, must land which is owned by every United States citizen be scarred for the sake of the few who will benefit financially by its exploitation?"

Boyd Norton of Idaho Falls doubts that even Idaho would profit. "I'm sure the metal is worth what they say, but the profit is made by a New York corporation and Idaho is left with a mess to clean up."

How do you balance the short-term profit for the few against the long-term for the many? There will never be more wilderness than there is now, there will never be more land. And scenic beauty grows in value as it becomes scarcer and man's handiwork compounds the destruction and the ugliness.

Prophets are usually without honor and proof is not often available in neat packages. But an anguished letter written to *American Forests* (December 1969) by Eldon L. Erickson of Duluth, Minnesota, projects White Clouds into the future clearly enough.

> . . . the fear and concern of the conservationists [for White Clouds] are fully justified and with firm footing based on actual proof as to locale—namely the Mesaba Range of northeastern Minnesota.
>
> Locate on a map these once prosperous mining towns—Chisholm, Eveleth, Mountain Iron, Virginia, Hibbing Coloraine, and then down toward Grand Rapids. This region, comprising a distance of over 40 miles, is a continual panorama of closed and abandoned high-grade iron ore mines—a reminder of a highly productive past which built these towns, induced the immigrant to live and work here, brought high production and huge profits for the down-east steel mills and provided a tax base for the local region.
>
> Finally the high-grade iron ore supply was depleted because of overproduction to meet the needs of both peace and war. The only reminder we have left of the prosperous past is the mine dump. What a pitiful sight! A dirty conglomerate of weeds, abandoned railroad tracks and switch houses, cable and wire strewn everywhere. Yes, a lot of this happens when a large mining effort closes down.
>
> Don't disregard the concern of the conservationists; there is adequate

proof that it can happen and will happen unless the conservationist and the preservationist makes his voice heard effectively.

One more item I'd like to mention in passing. A certain mining company on the shores of Lake Superior has a novel and most effective method of disposing of its mining throw-away—in the waters of Lake Superior. This can either make a person cry with remorse or roar with rage. I'm doing both because I will not sit by and let this happen. My duty is to help stop this despicable insult. God help me if I fail.

Here, then, is the dilemma of multiple use. The wardens of the public lands are obligated to consider all values equally in the public interest. But in the case of East Meadow Creek a single value—timber—became the dominant value. At Castle Peak a single value—mining—became the dominant value, outweighing all the others. If multiple use becomes single use, who accepts the responsibility of sacrificing all the other values, the values of wilderness, recreation, scenic beauty, wildlife, clean air and water, and the intangibles of solitude and nature uncontaminated by man's slovenly technology?

What does multiple use really mean? If, in a forest area, one value—lumbering—supersedes the others, the timber crew in its operations has been given license to alter and damage the ecology, change the habitat of the wildlife and suppress the recreational aspects. In the case of East Meadow Creek it would have meant a permanent alteration of status, removing it from likely wilderness inclusion. In the case of Castle Peak, the destruction and pollution caused by open-pit mining would be even more serious.

Consider the implications of the 1872 law, still in force, which covers mining claims. This law permits anyone to enter public land and stake out a claim. If minerals are found on the claim, the prospector is authorized to buy the land from the government at $5 an acre if the minerals are in a vein or lode, or $2.50 an acre if they are in a placer deposit. A placer is an alluvial deposit from which the minerals are recovered by washing rather than by sinking a shaft to an underground vein.

Consider the background of this law. In 1872 the country had a total population of about 40,000,000, with fewer than 1,000,000 in the great reaches of the West. The government owned millions of acres of land that it had trouble giving away under the various homestead acts. So the wandering prospector was more than welcome to scratch for gold and to file a claim for the land if he had the nerve to brave the wilderness. In 1872 Custer hadn't yet come to the Little Big Horn, the Indians were gathering their forces for a last-ditch attempt to convince the U.S. government it should keep at least one of its treaties made with them, and the Apache or the Sioux were likely to cut a prospector's career short anyway. It seemed a little picayune to begrudge him the land.

A hundred years later, with the population getting to the 300,-000,000 mark, the wild country almost gone and nobody giving away any land to speak of, the lone prospector and his burro replaced by the mining syndicate staffed with college-trained geologists and metallurgists prospecting by helicopter and Geiger counter, the mining industry would like to carry on as though this was 1872 and the land was still limitless.

Mike Frome, one of the best of today's conservation reporters, says that even the intent of the 1872 law has been perverted. Where the old-time prospector looked for gold, silver, copper or lead, mining claims in the fifties were filed by the thousands for sand, gravel or building stone.

This meant not just a few men sinking a shaft in a hillside to take out a few tons of rock in the search for gold, but massive machinery digging out millions of tons of sand and gravel.

> Evidence presented to the Senate [says Frome in *American Forests*, December 1969] showed that in a three-year period claims rose in Arizona by 700 per cent, and in New Mexico and Utah by 400 per cent. In 1955, new claims were being filed on the National Forests at the rate of seven every hour, or 5,000 every month. Throngs of the new miners had such noble purposes in mind as obtaining land for speculation, real estate development, tourist resorts, summer cabins, filling stations and timber. One modern miner patented his sand and gravel claim at the edge of a growing city for $2.50 an acre, then sold it soon after for $2,500 an acre—quite a reward for his initiative.

In cattle country, multiple use somehow gets mixed up with grazing rights for ranchers. The Bureau of Land Management conducted hearings in Arizona in 1966 to retain 31,000 acres in public ownership for multiple-use management. Governor Jack Williams appealed the decision for retention and claimed the land for the state, but was overruled. With the advent of the Nixon Administration the governor tried again. His motive, apparently, was to make the land available to the cattlemen, for he turned down an offer by the Bureau of

Land Management for an exchange of other Federal land in another part of the state. The new administration overruled the Bureau of Land Management to give the governor what he wanted, leading to a front-page article in the Arizona *Republic,* which stated angrily that the governor and the Nixon Administration were benefiting private individuals trying to avoid Federal land-management policies and allowing them to use the same land under a state administration with lower grazing fees, and without the multiple-use or land-improvement qualifications imposed by Federal regulations.

This is the same Governor Williams who, in 1967, at the height of the struggle over the Grand Canyon dams, replied to conservationists who asked him to intervene that he was "dismayed at those who write me urging I save it [the Grand Canyon] from destruction. I know of no evil forces," said the governor, "about to ruin this magnificent work of God."

In a society of such illusory values, it is consoling to reflect that the poets, after all, are the true realists. "In wildness is the preservation of the world," wrote Thoreau. And Carl Sandburg wrote: "There is an eagle in me and a mockingbird—and the eagle flies among the mountains of my dreams and the mockingbird sings before the dew is gone from the Ozark foothills. And I got the eagle and the mockingbird from the wilderness."

John Muir, who founded the Sierra Club in 1892, loved the Sierra Nevada with such intensity that his friendship with Gifford Pinchot, another pioneer conservationist, fell apart when Pinchot backed the construction of the Hetch Hetchy Reservoir for San Francisco in Yosemite National Park, which Muir had almost single-handedly created.

Testifying at the hearings on the reservoir, Congressman William Kent had said, "I hope you will not take my friend Muir seriously, for he is a man entirely without social sense." Which remark, of course, raises the question, whose social sense? Muir knew San Francisco had to have water; he just didn't see the necessity of putting the reservoir in the middle of Yosemite Park, a region so incredibly beautiful that it literally defies description by word or picture.

There is another point of view. Mortimer B. Doyle of the National Forest Products Association told the Portland, Oregon, Chamber of Commerce in March 1969:

Tragically, preservationists, unrestrained by logic or personal property considerations or community responsibility, wage effective warfare to achieve their ends. They persuade politicians, they persuade public officials, they persuade editors and the intellectual community to support their causes . . . and they win a discouraging number of times. It is my candid opinion that the entire population must resist this over-commitment of our nation to the "locking up of lands psychology" or we shall be denied the flexibility of resource management now assured through the concept of multiple use for the benefit of all.

This quotation goes along with another remark about preservationists made at a multiple-use meeting: "The greed of these people is monumental."

It is always instructive to listen to a man who wants to make a profit out of something calling a man who doesn't names. And it is also instructive to realize that Mr. Doyle says "multiple use" and thinks single use—wood. The "flexibility of resource management" he refers to has been so successful in the hands of the forest-products industries that it brought on the attempted raid of public forests in which Mr. Doyle's organization participated.

"The rate and extent of plunder," said Michael Nader in *The Living Wilderness* * of Autumn 1967, "far exceed the rate and extent of sanctuary. Yet it is a strange paradox that makers of violence to the land become stentorian about the 'greediness' of the non-plunderers. Our employed custodians too often find themselves, whether in good faith or by persuasion, serving the despoilers."

How far, then, is it necessary to go to satisfy the single-use demands of the multiple-use advocates? Ex-Secretary of the Interior Udall wrote in "The National Parks of America":

Because some segments of the public clamor for extensive road systems in parks and wilderness areas is not sufficient justification for uninhibited development. Because some people cannot walk or climb, or will not do so, does not justify building a road to every scenic overlook. Because some people like to see wilderness from the veranda of a modern hotel is not sufficient justification for building hotels within national parks when their location outside a park would provide necessary accommodations without encroaching on the natural scene. A wilderness trampled by thousands of refugees from the city is no longer a wilderness and the only way it can be maintained in its natural state as the population increases is to keep people out—to limit access. You would make reservations and

* The magazine of The Wilderness Society.

wait your turn—it would be as simple as that. Park and wilderness rationing in this country is not merely a prospect for the remote future, but could conceivably become necessary in the years ahead.

An hour's ride north of New York City is Bear Mountain Park in the Highlands of the Hudson. The area is dotted with lakes and laced with foot trails. Since the park was established the traces of old farms have all but vanished and the forest and wildlife have come back. But the population explosion has come and stayed.

On a Sunday in summer the cars clog the roads and the buses and Day Line boats disgorge thousands upon thousands from the city. A great sweep of meadow beside the Bear Mountain Inn has been converted to parking space, and new parking lots have been built on the slope below, but the overflow spills out all along the roads.

On the shore of Hessian Lake, which reflects the steep cliffs of the mountain, an amusement park has been built. There are screeching rides and shooting galleries and blaring jukebox noises and the smell of hot dogs and popcorn, and the ground is enriched with the debris of fun—paper plates and cups and discarded cans and bottles.

A few hundred yards away, where the ski slope begins, there are trails to take and mountains to climb. If you are of a mind you can in minutes leave the noise of jukebox and transistor radios behind and lose yourself in woods that are wilder now than when George Washington marched his troops behind these hills to conceal them from the British gunboats on the river.

But not many do. The bulk of the crowd stays here, near the great, timbered, still lovely inn, in the security of noise and the friendly congestion of the crowd. They do not really want to leave the city behind. They bring the city with them, and for what they want, Coney Island or Disneyland would do exactly as well. They do not want or need the hills and sky, much less the woods, where goblins lurk. But a benevolent park administration has brought the city out for them and instead of propagating the unique values of the park has set the lowest standard of public taste as the official value.

So in this park, if you look now for a quiet trail along Beechy Bottom, where once on an afternoon you might see a dozen deer and a quick fox or more leisurely raccoon, you will no longer find it. There's a paved parking lot bigger than a dozen football fields and an immense picnic area with tables and fireplaces and refreshment stands, and nothing that remotely resembles the lost and lovely valley that was.

Still, what's to be done for the thousands who come from the city, whose idea of wilderness is indeed a picnic area comfortably close to the parked car, where transistor radios blare and litter baskets overflow? Mustn't we make provision for them?

We must. And, most critically, we must in the Northeast, which, with a quarter of the population, has only 4 percent of the park and forest acreage of the country. There is still open space, says recreationist Laurance Rockefeller. "But we have to bring it closer to where the people are, bring the symbols of nature back into the daily life of the city."

The debate breaks down into some fairly definable issues. The remote wilderness is not for those who cannot, or will not, walk, ride horses or bicycles, or who insist upon meeting nature from the seat of a motor vehicle. Wilderness purists resist the building of more roads into parks or primitive areas on the grounds that it negates the very wildness supposed to be preserved, and damages the ecology.

Improved campsites, with sanitary facilities and conveniences, are very popular and each summer are so jammed with cars and trailers and tents that a new kind of slum has been created, making the goal of solitude and quiet a bad joke. Millions of cars create monumental traffic jams and much pollution on the roads, which is self-defeating. Wildlife finds it difficult to survive this onslaught of people and machines, which overrun and trample park and forest areas.

Timber operators and mining interests that reap the benefit of special privilege for a few from land that belongs to all resist the inclusion of lands in park or wilderness areas as though it belonged to them. In some cases their operations are marginal and the land might better be left in wilderness.

And always the proponents of development offer a popular cliché: Parks are for people—meaning not for trees and animals and birds. Let's grant it. The birds and animals and trees might not agree, but people are in control, so parks are for people. But do people come to the parks to see people or to see birds and animals and trees? Abuse the park, drive out the birds and animals, and what is left?

A reasonable compromise would seem to be a program for developing recreational areas within and close by cities for those who have no need of the wilds and who want the type of recreation typified by bathing areas, tennis courts, picnic grounds and amusement parks. The development of mass rapid transit, which we badly need anyway, would enable large numbers of people to reach these places easily and inexpensively and would take great numbers of cars off the road. Ski areas are reachable by special ski trains and buses, eliminating the need for individual cars.

As for wilderness users—their needs are altogether different. They do not bring the city with them; they are looking for a totally different experience.

An interesting study was performed in the summer of 1965 by William R. Catton, Jr., professor of sociology at the University of Washington in Seattle, with John C. Hendee, forest researcher at the Pacific Northwest Forest and Range Experiment Station. They wanted to find out what people who used the wilderness were like, and if they were different from most people. They prepared a detailed questionnaire containing 53 statements with which the recipients were asked to agree or disagree or comment on. These questionnaires were sent to 2,000 people who had visited the wilderness areas of Glacier Peak in the Cascade Range of Washington, the Three Sisters Wilderness in the Oregon Cascades, and Eagle Gap Wilderness in Oregon's Wallowa-Whitman National Forest.

In spite of the length of the questionnaire, 71 percent of those circulated were completed and returned with detailed comments that produced two immediate conclusions. One, these people had strong, definite opinions; two, they were well educated and articulate.

In fact, more than 30 percent were found to have postgraduate educations, and more than 60 percent were in the top 10 percent of the population in terms of formal education. Most were in the age group between 19 and 55 years; three-quarters were married, and of these, 85 percent had children.

Most were experienced wilderness voyageurs, with 70 percent having taken their first such trip before the age of 15. About half said they were often accompanied by three or more of their closest friends, and about the same proportion said their trips were usually family affairs.

Although they liked wilderness and solitude, these people were definitely not loners, not misanthropic or antisocial. Few went into the wilderness alone. To the contrary, the typical group was a small one seeking a renewal of family bonds or friendship through close association under intimate conditions and as a contrast to the impersonal contacts of civilization and its material or self-seeking aspects.

Few—10 percent or fewer—went on organized group tours, preferring to arrange their own, and reflecting individuality and a lack of conformity. Fewer than 30 percent belonged to organized groups, even conservation or outdoor clubs. Those who did belong to conservation organizations were more likely to live in urban areas but also made more wilderness trips, were somewhat better educated, had more friends interested in wilderness, and were more preservationist about maintaining the wilderness.

The conclusions reached by Professor Catton and Mr. Hendee were that, contrary to what one might have expected, wilderness trips are true social behavior, and wilderness users have high I.Q.s and good educations, are strong individualists, responsible and sensitive but not easily coerced or ordered, who will accept controls only when they understand the reasons and the need for them.

The no-compromise attitude of wilderness purists is better understood in the light of continual attempts to subvert established wilderness areas under the guise of multiple use. For example, the Adirondack Forest Preserve was established in 1885, and in 1894 a provision was added to the state constitution which read: "The lands of the State, now owned or hereafter acquired, constituting the Forest Preserve as now fixed by law, shall be forever kept as wild forest lands. They shall not be leased, sold or exchanged, or be taken by any corporation, public or private, nor shall the timber thereon be sold, removed or destroyed."

"Forever wild" became a catchphrase or slogan, but in spite of it there are continual attempts to change or amend the provision. In fact, it has been amended to accommodate the building of major highways, ski trails and reservoirs.

Attempts have been made, and continue to be made, to permit lumbering and construction of summer cabins and power dams. The New York State Conservation Department appears to take the forever-wild dictum with more than a grain of reluctance.

Not long ago it reported that this policy was creating a "mature" forest with a reduced capacity for supporting the deer. The department proposed, therefore, "habitat improvement" operations on 150,000 acres of land designated as "deer winter range" and, because of the presence of roads or dwellings, considered nonwilderness areas. The improvement was to consist of cutting or killing 30 to 500 trees per acre over a period of years in order to open up the forest and provide more browse for the deer. The trees so cut would be sold to defray the cost of the program.

A suspicious mind might conclude that this was only a disguise for a lumbering program in the State Forest Preserve. It should also be fairly obvious that if the number of deer exceeds the carrying capacity of the range, improving the range will only result in an increase in the number of deer, and shortly the situation will be right back where it started.

What the Conservation Department was doing, intentionally or not, was in effect trying to cut up the Forest Preserve into sections—about one-third held as wilderness and the other two-thirds subject to management, improvement, development, lumbering and road building.

Urging Congress to expand the national park system in the sixties, Secretary Udall said: "In 1940, 130,000,000 Americans had a spacious National Park System of 22,000,000 acres; twenty years later, a population which had grown to a more mobile 183,000,000 inherited an overcrowded system that had been enlarged by only a few acres."

Secretary Udall managed to add four new national parks and several smaller areas—national monuments, recreation areas, seashores, and so on to the system, but even so it lagged behind the demand. Visits to the parks increased ten times faster than the population gain.

The Wilderness Preservation Act, signed by President Johnson in 1964, has made the preservation of at least some of our wild heritage a matter of national policy, but it is a policy continually on the defensive. Under the terms of the act, the Department of the Interior is obligated to review every roadless area of 5,000 acres or more and every roadless "island" within the National Wildlife Refuge System, to determine their suitability for inclusion in the Wilderness Preservation System. Public hearings are called regularly, but unless you happen to be a member of the Wilderness Society or interested local organization, you are not likely to hear of them.

Not all wilderness is in remote areas of cloud-piercing mountains.

Early in 1966 a call went out from the Wilderness Society to its members to participate in a hearing held in Morris County, New Jersey. The hearing was on a proposal to make the Great Swamp, right on the doorstep of the biggest metropolitan area in the country, a national wilderness area.

There was urgency in this call, as there is in every such call to a hearing. Since 1959 the Port of New York Authority had been trying to acquire the Great Swamp for a metropolitan jetport. The first effort was blunted by a spontaneous uprising of the local citizenry, although it must be recorded that many who reacted so violently against the Port Authority did so out of fear that the scream of jets overhead would send real-estate values plummeting rather than out of interest or real understanding of what the Great Swamp really is.

Swamps have always had a bad reputation. People think of them as treacherous bogs where fever-laden mists veil ghostly lagoons hiding fearsome creatures, and quicksand waits to engulf the unwary.

There are marshes in the Great Swamp, but there are also lovely meadows and stretches of virgin timber with oak trees that measure 14 feet in circumference and beech trees with magnificent global crowns. There are laurel bushes 15 feet high and wild orchids and thick beds of ferns and azalea and rhododendron.

The ponds are host to wild ducks and geese and herons and bitterns. There are deer and beaver and pheasants and raccoons and otters and mink and fox and muskrat. There are more than 1,000 different kinds of plants in the Great Swamp and at least 178 species of birds. All this great cornucopia of life—right on the doorstep of New York City.

About 20,000 years ago the whole area was a large lake, 30 miles long and 10 miles wide—Lake Passaic—created by the retreat of the Wisconsin glacier. The lake spilled over the Little Falls precipice at Paterson and drained away to the sea, leaving a boggy area punctuated by islands of higher ground covered with magnificent hardwoods—chestnuts, oaks, hickories, maples and beeches. The great sponge of the marsh still serves as the headwaters of the Passaic River.

William Penn bought the Great Swamp from the Delaware Indians in 1667. "Bought" is a euphemism; the Indians had no idea they were actually selling land, because they had no concept of ownership and even less of what the paleface was talking about. At any rate, the tract Penn "bought" then measured about seven miles by three. Even now, with its borders nibbled away and its trees cut to make charcoal for the old iron foundries, about 8,000 acres are left in forest, marsh and meadow. To local naturalists the Great Swamp is a place of incredible riches. It became a field classroom for students in elementary school and a field laboratory for college students from Drew University, the College of St. Elizabeth, and Fairleigh Dickinson University.

When the first threat of a jetport appeared in 1959 a campaign to raise money brought in more than a million dollars from thousands of people, and with this money the Great Swamp Committee bought more than 3,000 acres of land and donated 2,700 acres to the U.S. Fish & Wildlife Service for a wildlife refuge. This was the nucleus for a projected 6,000-acre sanctuary.

In spite of the existence of a wildlife refuge, the Port Authority made another bid for the Great Swamp in December 1966. But in the meantime the Department of the Interior had made the requisite study under the Wilderness Act to see if the area merited inclusion in the Wilderness System. In the fall of 1968 President Johnson put his signature on a bill creating a 3,750-acre Great Swamp wilderness and removing it (barring reversal by a future Congress) from the grasp of the Port Authority permanently.

At the moment, then, there seems no current danger to the Great Swamp. If the Port Authority is looking elsewhere for a jetport site, the fact has not been publicized. There are few places in the metropolitan area that would free the Port Authority from problems of jet noise. There is no indication that the Authority has given consideration to the theoretical possibility of building a floating airport off the coast. With such an airport, all takeoffs and approaches would be over water and need never come near any inhabited area. As for pollution of the sea, this problem must in the long run be solved anyway by changes in engines and fuel.

If the Great Swamp story is one of limited yet spectacular success, the attempt to save another patch of wilderness, the Big Thicket of Texas, is one of frustration. A bill sponsored by Senator Yarborough of Texas to preserve 75,000 acres—about 2 percent of the area—has been stalled for years; meanwhile the Big Thicket is being systematically cut to pieces.

The area originally comprised about 3,500,000 acres, and had the reputation of a dense, impenetrable jungle. The tangled undergrowth that baffled early travelers, however, was for the most part along stream banks. Elsewhere were sandy slopes with groves of longleaf pine and great beech forests. Its location makes the region a point where three climatic zones merge—south, west and the subtropics.

Palmetto palms and wild orchids—20 varieties—grow beside subtropical bayous. Nearby are the bald cypress, magnolia, sweet gum and tupelo of the Midsouth, while not far away grow mesquite, yucca and cactus, the plants of the western desert. There are even areas of Appalachian flora. Botanists have said this may be a region of "critical speciation," where new species evolve in response to the pressures of the environment.

There is a blend of animal life too. The expected mink, otter, bear, raccoon and opossum are joined by jaguars and ocelots from Mexico, and armadillos and civet cats, as well as alligators, lynx, wolves and flying squirrels. The Big Thicket is a marshaling point for migratory birds and a rookery for such magnificent species as the snowy egret, many kinds of herons and the roseate spoonbill.

So large a forest could hardly escape the axe. Settlers began moving in and cutting the trees early. At the time of the Texas war of independence much wilderness still remained and even some Indian tribes. Sam Houston made plans to hide his Texas rebels in the Big Thicket if the Mexican Army defeated him, and similar ideas occurred to outlaws of every stripe. The Big Thicket became the Texas equivalent of the Hole-in-the-Wall country farther north, a hideout for men fleeing the law.

Although they were southerners, Sam Houston and many of his followers were opposed to slavery and unsympathetic to the Confederacy. When the Civil War broke out, many of them refused to serve in the Confederate Army. They hid out in the Big Thicket and were pursued by Confederate soldiers looking for them. Skirmishes between the deserters—jayhawkers, they were called—and Confederate soldiers continued through the war and bestowed picturesque names on the scene of many little battles: Deserter's Island, Panther's Den, Doc Trull Hammock, and Honey Island, so named because deserters here swapped wild honey for gunpowder and tobacco.

Over the years the Big Thicket has been whittled down by about

50 acres a day, and now less than a tenth remains—300,000 acres or so—much of it cutover and second growth. For this reason the Park Service has argued that it is too late, there isn't the minimum 5,000 acres of roadless or otherwise unaltered land that would make it eligible for wilderness classification. But the Wilderness Society, the Sierra Club and the Audubon Society believe the forest would come back if it were saved now. The Audubon Society is particularly interested because the Big Thicket has been found to shelter two birds formerly thought surely extinct, the whooping crane and the ivory-billed woodpecker.

The greatest menace to the Big Thicket now is from the lumber companies. That supposed spokesman of rugged individualism, *The Wall Street Journal,* reported on July 1, 1968:

> Because pine is more profitable than hardwoods, these companies are systematically eradicating the hardwoods. Grand old magnolia trees—like beech and ancient oaks a nuisance when growing in pine-producing areas —are felled and cut into railroad ties. One smaller lumberman sprayed his hardwoods with defoliants, wiping out an entire rookery of hundreds of herons, egrets and their young. "It is obvious that the many direct and indirect encroachments, if allowed to continue, can only result in the complete destruction of the native forest," says the Park Service.

The Wilderness Society reports that the practice of spraying defoliants is continuing by airplane, killing trees, birds and other animals. A huge magnolia tree, a thousand years old, was found drilled and injected with arsenate of lead, a deliberate vandalism intended to reduce the attractiveness and wilderness potential of the woodland.

The newspaper quotes G. W. Stanley, a vice-president of the Kirby Lumber Company, as agreeing that hardwoods must be eliminated where pine can be grown profitably. "To the layman who refuses to inform himself as to what good management is, it looks like a terrible hodge-podge," said Mr. Stanley, and went on to state that Kirby's main responsibility is to make a profit, to sustain a multimillion-dollar plant at Silsbee, Texas, and to supply supplemental materials to the various associated industries. As to giving up any land, even for a relatively small national monument, said Mr. Stanley, that is "sort of like asking someone if they can get by without their little finger."

Kirby has been a center of controversy in the Big Thicket since the turn of the century, says the paper, because it acquired some of its present landholdings through title suits against previous owners.

In 1965 a Big Thicket Association was formed which urged Governor John Connally to take action for a state park. Governor Connally expressed himself interested and paid an official visit to the Big Thicket. He arrived, according to *The Wall Street Journal*, in a private plane owned by Eastex, Inc., a major timber owner of the locality (and a subsidiary of Time, Inc.). Curiously enough, following this visit, the proposal for a state park no longer seemed to engage the governor's interest.

Governor Connally did appoint a three-member State Park and Wildlife Commission. The members included a road contractor, an oil and gas producer, and the president of a brewery, none of whom had ever distinguished himself by any expression of sympathy for conservation.

A year after this Justice William O. Douglas visited the Big Thicket and found that "the scene was one of devastation. A major hardwood cutting project was under way; magnolias had even been cut on a public road right-of-way. 'In the road's edge dozens of magnolias lay freshly cut,' the Justice wrote. '. . . they were not cut for flooring, for paneling, or for railroad ties. They were cut for sheer destruction and the trunks lay rotting.' "

Lance Rosier of the Big Thicket Association tells of visiting a rookery close by an area where the chain saws were cutting trees. "There were hundreds of shotgun shells lying around," he said. "You could just pick them up." The vandals were not even concerned enough to remove the evidence.

The Yarborough bill is still stalled in Congress for "study." "You can study this to death," says Dempsie Herley, president of the Big Thicket Association. "In five years there will be no Big Thicket." Only a requiem.

To illustrate the abrasions of multiple use there could hardly be a better example than the clash between preservationists and recreationists in 1969 over Disney's Mineral King project in the Sierra Nevada of California.

Mineral King is an alpine valley, 7,800 feet high, a perfect snow bowl surrounded by spectacular frosted peaks, some of them 12,000 feet high. Geographically and topographically it is part and parcel of Sequoia National Park and is surrounded on three sides by the park, but was gerrymandered out of the park because of mining operations in the valley about 1879. The mines petered out. Nevertheless when Sequoia National Park was established in 1890, Mineral King Valley was cut out and became part of the forest preserve and a wildlife refuge, administered by the Forest Service rather than the National Park Service.

The valley was a natural prospect for a ski resort, and preservationist groups such as the Sierra Club did not oppose development of ski facilities. The Sierra Club assumed that a modest ski development would do little harm to the landscape or the ecology. Its policy does not oppose well-planned recreational development.

In 1949 the Forest Service invited bids from developers to create a ski resort in Mineral King Valley. There were no takers, presumably because of the difficulty of access. The single dirt road into the valley was an impossible wagon trail, steep and narrow, with switchbacks and drop-offs. From Three Rivers, at the edge of the San Joaquin Valley, to Mineral King, the road climbed 6,500 feet. Moreover, this access road crossed Sequoia National Park, which required an easement.

By 1965, when the Forest Service invited bids for the second time, two changes had become apparent. On the one hand, human activity in Mineral King Valley had further diminished to a few fishermen and summer campers, and the region had largely reverted to wilderness. On the other hand, ski fever had hit California and developers were looking avidly for new sites.

This time six developers responded to the Forest Service invitation. One bid was approved. Walt Disney Enterprises was granted a three-year permit to develop a master plan for Mineral King. Early in 1967 the shape of the Disney plan appeared and promptly ran into its first roadblock.

Secretary of the Interior Udall turned down the Forest Service request for permission to construct a new highway through Sequoia National Park to Mineral King. The action precipitated a major clash between Interior and Agriculture. And the rumblings of a growing opposition began to be felt from three groups—the Sierra Club, the Wilderness Society and the National Parks Association.

It should be noted here that the Disney organization has compiled an enviable record in conservation. It has received 37 awards for its contributions, mainly its nature films. Walt Disney was a life member of the Izaak Walton League and an honorary life member of the Sierra Club. Nonetheless, as early as 1967 both the Sierra Club and the National Parks Association had come to the conclusion that the Disney project was flamboyant rather than well conceived, and that it would be destructive of too large an area, considering the fragile ecology of these high altitudes.

The American Forestry Association, after sitting on the fence for a couple of years, decided in the spring of 1969 that the Disney plan was a "golden opportunity" for multiple use. "Sometimes," said William E. Towell, vice-president of the organization, "I lose patience with fellow conservationists who are out to 'save' everything." Mineral King, he suggested, could serve as an invaluable experiment to demonstrate public use and enjoyment and still protect the ecological values. National parks are being ruined by hordes of visitors and cars because of lack of planning. Mineral King could show the way—why not let Disney do it?

The National Parks Association suggested that the Disney organization was an ideal one to create an entirely different kind of area in Mineral King, a living museum where visitors would be able to observe and photograph wild animals in their native environment. The Sierra Club dismissed the whole project as too big, too destructive and too expensive for the taxpayer, who would be asked to pay for a new 20-mile access road at more than $1,000,000 a mile.

Ignoring the rising tide of objection, the Forest Service accepted the Disney plan with enthusiasm, and its parent bureau, Agriculture, increased the pressure on Interior. By 1969 Secretary Udall gave in and granted permission for the access road across Sequoia National Park.

The Disney organization had meanwhile spent more than $500,000 in surveys, employing engineers, hydrologists, architects, foresters, meteorologists, soil scientists, resort operators and ski experts. Snow-survey teams had spent four years in the valley mapping snowfalls and charting avalanches. Disney experts visited 22 ski resorts in Europe and came back with bulging notebooks, convinced they could create a much more efficient and comfortable ski resort.

The Disney plan called for the construction of a year-round Alpine village—a twentieth-century projection of a Swiss or Tyrolean picture-postcard town—at a cost of $35,300,000. The village, large enough to handle a million visitors a year, would contain two five-story hotels, a 500-room dormitory, 1,200 cabins, 10 restaurants, an auto-reception center, a five-acre, nine-level covered parking lot a mile from the village and serviced by an electric cog railway, an auto service station, horse corrals, skating rinks, heated swimming pools, a theater, a chapel, and a gondola-lift center. Twenty-two ski lifts would rise to the surrounding mountains some 4,000 feet above. An enclosed lift would carry diners to a restaurant part way up one peak, or, if they wanted still more altitude, would go on to another restaurant at the top. As many as 3,300 beds would be available for overnight visitors, and the slopes would accommodate 8,000 skiers at a time. The Sierra Club estimated that as many as 16,000 people could be in the valley at once, making a population density of 53 persons to the acre, or 34,000 per square mile—a greater density than New York City.

The new access road would be designed as a two-lane highway, with occasional third-lane stretches for passing, and geared to speeds up to 50 miles an hour (although the Clarkeson Engineering Company, retained by the Park Service, declared the road unsafe at these speeds).

The Forest Service saw the project as an ideal recreational use of the land and estimated fees from the Disney organization at $300,000 the first year, doubling in five years. "We're happier with Disney than a frog in a mud puddle," said Jim James, the forest supervisor.

The project had other, formidable support. Senator George Murphy, Governor Ronald Reagan and past-Governor Pat Brown, civic groups, some local newspapers and most local businessmen saw a happy influx of tourists and an overflow from Mineral King to the San Joaquin Valley. Visalia, the capital of Tulare County, projected $500,000,000 in taxes and payrolls within 10 years. Ex-Senator Thomas H. Kuchel of California, a sincere conservationist, said, "If we fail to develop selected areas such as Mineral King, the 50,000,000 people who will be in California before the end of this century will spill over the sides of the coastal cities and ravage the Sierra with unplanned and undirected enthusiasm for the vanishing outdoors."

The Disney organization was offended at the intimation that it was planning to erect a tasteless, overblown mountain Disneyland and that it was insensitive to ecological damage. Robert B. Hicks, project manager for Disney, an economist and engineer and a lifetime skier, said, "It is unfair to think that Disney did a good job at Disneyland, but will do a bad job elsewhere. Why not assume this project will be just as appropriate for Mineral King as Disneyland was for the recreation needs of Anaheim?"

The Forest Service insisted there would be no pollution, no erosion, no damage to the giant sequoia trees through which the new access road would go, and no alienation of the public trust. It brushed aside the objection that Mineral King is a wildlife refuge and that a million visitors a year might adversely affect the status of such endangered species as the cougar, grizzly bear or even rarer California condor. Its serenity was firmly rooted in multiple-use philosophy, through which the Forest Service assigns to each area under its jurisdiction one or more uses for which it seems suited. Of the five such uses—watershed management, grazing, lumbering or mining, wildlife and recreation—Mineral King seemed best suited for recreation. It was immaterial to insist that this choice, as had other choices, simply wiped out all the remaining uses. It was immaterial to say this is single use; the Forest Service was not concerned with this kind of hairsplitting.

A stream that cuts through Mineral King Valley—Monarch Creek—is a kind of casual stream, carving its way and changing channels at will. Such indecisiveness might one day take it through the lobby of one of the new hotels, so the Disney plans called for cutting a permanent channel for the creek and putting it in there to stay.

This is exactly the kind of thing that raises the hackles of conservationists. "That's typical of the Forest Service," said Michael McCloskey, conservation director of the Sierra Club. "Nature doesn't know what it's doing. We know better."

A real concern of the conservation groups was that a successful Disney project would encourage other developments on the fringes of the area, as indeed "honky-tonks and beer parlors" have grown up around Disneyland in Anaheim. One such proposal, the Seaborn-Wells project, five miles west of Mineral King, was approved early in 1969 by Tulare County. This project called for a 30-unit motel, a 72-unit lodge, 20 condominiums, 48 one-family units, a gasoline station, a swimming pool, stores and a restaurant.

The situation has its comic aspects. The Forest Service immediately opposed this development, but now finds itself in a semantic dilemma. It is opposed to a private development on private land, yet endorses a private development on public land. Forest Service influence helped Tulare County officials decide to turn down the Seaborn-Wells application, but the company will appeal. If it wins the appeal, the Forest Service intends to have the land condemned, with the prospect that it may now cost $2,000,000, whereas before the Disney bid in 1965 it could have been bought at almost any price.

Another new plan for a development has been announced at Three Rivers, 29 miles away, for a 12,000-acre resort, bearing out the misgivings of conservationists that Disney fever would spread.

Against all the high-priced talent and financial power supporting the Disney project, the Sierra Club calmly filed suit in the U.S. District Court on June 5, 1969. The suit named as defendants Walter J. Hickel, Secretary of the Interior; Clifford M. Hardin, Secretary of Agriculture; John S. McLaughlin, Superintendent of Sequoia National Park; J. W. Deinema, Regional Forester; and M. R. James, Forest Supervisor of Sequoia National Forest. The suit charged that these officials acted illegally in granting permission to Disney Enterprises to develop a commercial resort at Mineral King.

The basis for the charge of illegality is in four parts:

1. The law clearly states that the Forest Service may permit private development on public land in an area no greater than 80 acres. At Mineral King the Forest Service is giving Disney access to 13,000 acres. The valley itself contains about 15,000 acres, of which 400 acres would be taken up by buildings or would otherwise be bulldozed and altered, and 13,000 acres would be affected by ski runs, trails and ski lifts. The towers at the top of the ski lifts would invade Sequoia National Park at a number of points around the rim.
2. The Forest Service has overreached its jurisdiction at Mineral King. Responsibility for wildlife in a national game refuge was transferred by Congress to the Secretary of the Interior in 1939, removing it from Agriculture.
3. Development of Mineral King as a resort is in violation of its status as a wildlife refuge. The law says the Secretary of Agriculture may authorize other uses under multiple use only "so far as may be consistent with the purposes for which

said game refuge is established." A $35,300,000 resort cannot possibly be considered compatible with a wildlife refuge.

4. The Forest Service violated the law and its own rules in refusing to hold public hearings on the project before giving its approval to the Disney plan.

A second set of charges was leveled against the Park Service for yielding to pressure brought by the Department of Agriculture. These charges claimed that giving a commercial developer permission for a road to cross Sequoia National Park does not serve park purposes and, in fact, will damage the park. There are at least 103 giant sequoias below the proposed highway, of which 45 of the great trees are in a position of jeopardy from road construction.

Another claim is that the Park Service violated a Federal regulation calling for a public hearing on roads in national parks. In January 1969 the Interior Department published in the Federal Register policies for park roads, including an order to make public hearings obligatory. Secretary Hickel later repealed this order. The suit maintains that Hickel's repeal was illegal and that permission to proceed with the road is a violation.

"Who watches the watchmen?" asked *The New York Times* editorially on June 24, 1969. "The question is as old as Plato and as modern as the Mineral King case. Private greed, bureaucratic empire building and official irresponsibility are a recurrent threat to the common good. The Sierra Club is performing a public service in standing up to this threat."

Despite these apparently clear-cut violations of law, conservationists, while applauding the Sierra Club action, do not appear too confident of the outcome. There is too much money involved, too much influence. "Progress always beats Protectionism," says writer Arnold Hano writing in *The New York Times* Magazine.* And, noting that we have all suffered too long over Federal and state abuse of public land, that the first recommendation of President Nixon's new Environmental Quality Council was to suggest further drilling for oil in the Santa Barbara Channel, he concludes: "Perhaps then, the truest value of Mineral King is that it stands as the latest crass example of abuse of public land by the very agents to whom we entrust that land. One wonders how long America must tolerate such stewardship."

* "Protectionists vs. Recreationists," August 17, 1969.

The statement implies that the country is long-suffering and that the time will come when the people will arise in wrath against the representatives who have sold them out. Unfortunately the people as a whole couldn't care less.

"Parks are for people, not birds," makes good sense to most; the draining of marshes, the building of Disneylands in a wild valley are perfectly acceptable. Only when the rivers have become open sewers and the air sears the lungs and the trees are gone and the sun glares everywhere on concrete does the average citizen even begin to wonder what happened. Until then the few who saw it happening and who tried to stop it are abused as such men have been abused in all times. The man who says conservationists care more for birds than for people displays not only his ignorance but his own contempt for mankind, for life itself.

The fabled lumbermen of the North—the Paul Bunyans—are celebrated in song and story as American folk heroes. Who knows or cares that they clear-cut the Minnesota forests like a man shaving the stubble from his face and then dropped matches in the slash to leave 40,000 square miles of ruin behind them as they went off to Oregon to repeat the process?

Man is a strange creature. He talks endlessly of love but endlessly makes war and kills like no other species. Does the hunter who takes a high-powered rifle into the woods and shoots at anything that moves understand what he is doing—or care? Would he understand a man like Anthony Wayne Smith who says the trend toward death violates an ecological imperative—that men may not wantonly destroy the life around them under penalty of death for themselves? Or that the decision to inflict death upon another creature needlessly is a decision to move toward death oneself?

As life is an ecological imperative, wilderness is a cultural imperative, a fight against the encroaching concrete desert surrounding us, to Stewart Udall the "supreme human experience."

If, then, the protectionists are skeptical of multiple use, it is because they see so clearly that what is yielded cannot be regained and that too much has already been yielded. Wildlife, says Mike Frome, must be protected for reasons other than to enhance the pleasures of people. The disappearance of a species is a danger signal to the environment of which man is part. The loss of a bird of prey means not simply that we are deprived of the beauty of a hawk in flight, but that we have suffered the loss of an irreplaceable link in the ecological chain—the chain that connects us all.

How much wild country needs to be reshaped to sustain hordes of visitors in the name of recreation or multiple use? Says Martin Litton of the Sierra Club, "It would be laughable if it weren't so tragic, to hear people speaking of increasing the opportunities for recreation when they are wiping out the opportunities for the very highest and most ennobling kind of recreation, the contemplation of creation."

A strange report has come from overseas. It says that butterflies are disappearing from all over Europe. Butterflies in Europe and the last wilderness in America. We won't get many more warnings.

CHAPTER

12

Life in the Balance

THE director of Montana's Fish and Game Commission is in trouble with the governor for doing his job. From public statements made, it appears that Director Frank Dunkle's insistence upon environmental safeguards for new industry coming into Montana were an unpleasant surprise to the governor. Who ever expected the man to take his job so seriously?

"He's going to play his role and he's going to have to play it our way," Governor Anderson was quoted in *The New York Times* of July 27, 1970.

Dunkle has been guilty of several reprehensible actions. In 1963 he found Montana's blue grouse carrying dangerous concentrations of DDT and managed to get the Forest Service to suspend spraying of thousands of acres of spruce forest.

A little later he halted the feeding of elk with hay so contaminated by pesticides that it had been condemned for cattle feed.

In 1967 he initiated a court action to save a trout stream from contamination with effluent from the Butte copper mines.

In the fall of 1969 he issued a warning to hunters to avoid the state's pheasants as they were seriously contaminated with mercury from seed fungicides.

The same year he asked a lot of embarrassing questions: what effect new dams would have on the environment; how an open-pit copper mine in the Lincoln area would pollute the streams; how strip mines in the southeastern part of the state were to be restored; and what ecological effects might be expected from Chet Huntley's big new recreation area in Montana.

Mr. Dunkle made himself very unpopular in some quarters with all this questioning. His critics felt that the crises he saw were not as crucial as he thought and that what he was doing would discourage industrial development in a state high on scenery but low in population and income-producing occupations. A number of people, including his own five-man commission, urged Mr. Dunkle to resign. "I'm not a quitter," he is reported to have said. "You just don't understand the 20 years of my life I've invested in the department and the needs of the resources."

There are troubling aspects about this affair. How much government policing of the environment is political window dressing? How many Frank Dunkles are there—given a job but not expected to do it, and in trouble if they try to do it?

The opinion of those who are watching the problem is that this is a real issue. In spite of millions of words of political rhetoric and almost as many promises, little good has happened and a lot worse is on the way. The situation is more worrisome now than it was three years ago or five years ago. We move deeper into crisis, and it is not just American, it is world-wide. It is desperately serious in Europe, in the Soviet Union, in India and in Japan.

Only 14 percent of Italy's long coastline is relatively clean. The Baltic, an inland sea having a single outlet to the Atlantic, so that it flushes itself very slowly, has become a cesspool. All around it the countries of Germany, Denmark, Sweden, Poland, Finland and Russia dump industrial waste and sewage. DDT and mercury are accumulating, and oil films ride the surface. All that is needed to finish it off is one big tanker wreck, spilling a quarter of a million tons of oil. Such a castastrophe would destroy it, killing virtually all marine life and seriously affecting six nations with a population of 300,000,000 people.

The Rhine is more nauseatingly polluted than the Hudson, effluent making up 20 percent of its volume by the time it reaches the North Sea. The historic lake at Zurich is as dead as Erie. The fish are gone from the Tiber and the Arno in Italy, and the famed Venetian canals are so foul they have been called a "major cause of the decay threatening Venice."

The French are pleading with the Italians to help curb the pollution washing up on the Côte d'Azur and spoiling the fabled Riviera.

In Milan, smog has deducted three years from the average life span, while Tokyo's smog is worse than that of New York or Los Angeles.

Swedish forests are withering under a rain of sulphuric acid from the steel mills of the Ruhr. In Japan, Finland and Holland the harvests of shellfish are deadly with mercury. In Russia the largest body of fresh water in the world, Lake Baikal, is becoming discolored and foul with effluent from pulp mills. Detergents are being curbed in Canada. Singapore, still clean, has begun to take preventive action against a rising threat of pollution from cars and industry.

Around the globe, erosion, overgrazing and bad agricultural practices are creating new deserts. Pollution hangs choking umbrellas over the cities and the water is unfit to drink.

Meantime the population climbs. It is expected to double in 30 years to an uncomfortable total of 7,000,000,000, with a likelihood of famine in several parts of the world. Today's problems will look simple in the year 2000, and survival, perhaps already problematical, will have lost more of the shrinking odds.

For now the sense of crisis seems limited to the doomsday ecologists and the youth, and a coalition is taking place between environmentalists and the young. This coalition has discovered a new weapon—court action.

For years the conservationists employed the classic tactic of taking the issue to the public, of trying to arouse through education a sense of danger in what was happening. The result was failure after failure, with a harmful project long finished before any appreciable outrage could be measured in the public pulse. Failure and frustration led them to look for more effective methods, and they found it in the law. A legal injunction is fast, and it has the great merit of stopping the project while the arguments are being thrashed out.

This discovery gave teeth to the activities (and added hugely to the financial burdens) of the Audubon Society, the Wilderness Society, the Sierra Club, the Izaak Walton League and the others.

It gave birth to the militant Environmental Defense Fund with a slogan of "Sue the bastards!"

The basis of a growing number of lawsuits is that the public has a right to fresh air and clean water and some undestroyed landscape, and it doesn't require a special law to assert this right. Conservationists have been urged to use the Ninth Amendment to the Constitution, which says, "The enumeration in the Constitution of certain rights shall not be construed to deny or disparage others retained by the people."

If this seems vague, it has nevertheless already been used in pleading environmental cases, sometimes together with the Fourteenth Amendment. The Fourteenth Amendment says, "No state shall make or enforce any law which shall abridge the privileges or immunities of citizens of the United States, nor shall any state deprive any person of life, liberty or property without due process of law, nor deny to any person within its jurisdiction the equal protection of the laws."

Professor E. F. Roberts, of Cornell University Law School, considers that the right to free speech and to privacy covers the right to a decent environment, without which "the rest of our rights will prove illusory. We cannot enjoy our other rights if we are all dead."

It is true, says Professor Roberts in *Natural History Magazine*, August–September 1970, that this right has never before been articulated, "but until the advent of a potentially lethal technological society there has been no need to insist upon such a right."

The principle is being tested in dozens of lawsuits with the intention of halting individual projects detrimental to the environment and at the same time setting precedents in law that will be of use in following cases. The Sierra Club at this writing is supposed to have 55 suits in progress around the country. Some individual suits provide a hint of the more militant attitude that is now coming of age.

In Phoenix, Arizona, a pair of university professors and their wives are suing six copper companies for $2,000,000,000 on behalf of the 700,000 residents of Salt River Valley over air pollution they consider injurious to public health.

Carol Yannacone, wife of Victor Yannacone of the Environmental Defense Fund, has filed suit for $30,000,000,000 in damages from eight producers of DDT, "on behalf of all the people of the United States."

A similar suit was filed in California against the four major auto manufacturers, charging conspiracy to hinder the development of smog-control devices. The amount was not stated in the claim, but if damages were assessed on the basis of damage or injury claimed to be caused by exhaust pollution, it could also run into billions.

The Federal government is in the courts, too, with a suit against the Florida Power & Light Company for creating thermal pollution of Biscayne Bay; with another suit against Chevron Oil for its massive pollution of the Gulf of Mexico in the spring of 1970; and with suits against eight industrial companies in seven states for dumping mercury into lakes and rivers.

Realistically, there would seem to be small chance of a $30,-000,000,000 verdict, payable to the people of the United States, yet one large industry involved in environmental suits, United States Steel, says it is taking these cases very seriously.

To the conservation groups, which are also very serious about the suits, the money involved is nothing; it is both the establishment of the right to sue and the halting of an injurious activity that are important. The challenge to them is to prove injury as claimed.

"The courts," said Roderick Cameron of the Environmental Defense Fund, "want something a little more definite than people coming in and saying the sky isn't as blue as it used to be, or that the fumes are aggravating their sinuses."

The Environmental Defense Fund never sues for money damages. It asks only that the injurious action be stopped. It does not seek to close down a plant. It asks only that the offender do everything possible to stop the pollution. This may involve only the installation of "state of the art" pollution-control devices, which, of course may be only partly effective. But EDF is not trying to drive companies out of business or deprive local residents of jobs. It treads a narrow line on that dilemma, but zeroes in on offenders who refuse to make any effort at all to curb pollution.

Beginning its career with a hard-fought case on DDT in Wisconsin, the Environmental Defense Fund next took on the Federal government to see that the ban on persistent insecticides was actually enforced. In 1969 it sued the Army Corps of Engineers to halt construction of the Cross-Florida Barge Canal, which would have destroyed the Oklawaha River, and in 1970 stepped on Secretary Hickel's toes with a suit to stop construction of the Trans-Alaska oil pipeline, a threat to the fragile Alaskan tundra. That year EDF also initiated litigation before the Department of Health, Education and Welfare to immediately eliminate lead from gasoline.

EDF also plans to join Friends of the Earth in action to stop the SST. Litigation is also being considered to protect wetlands and estuaries, to stop predator-control programs and to check the killing of whales now facing extinction.

In September 1969 a national conference on "Law and the Environment" was held at Airlie House in Warrenton, Virginia. It was sponsored by the Conservation Foundation of Washington, D.C., and the Conservation and Research Foundation of New London, Connecticut. Participants were Ralph Nader, former Vermont Governor Phillip Hoff, Victor Yannacone, Congressman Paul McCloskey, Jr., of California, Phillip Berry and Mike McCloskey of the Sierra Club, and many others.

The conference agreed that conservationists were generally outgunned by industry, that expert witnesses were hard to rally, and that precedents in the law had to be established. They agreed on a fistful of needs:

An environmental law center
A new legal publication in environmental law
An early warning system to alert citizens to projects threatening the environment
An environmental ombudsman in Congress
Environmental law programs in law schools
Research service for environmental lawyers
Environmental law sections in bar association committees
More involvement generally by lawyers

At the conference Congressman McCloskey said:

I think that perhaps the true enemy of preservation of our environment is our system of government, and by that I mean that the local government and the county government which is entirely dependent upon the property tax and the increase of its payroll structure is the true enemy of conservation today. It may be that we must revise the entire structure of the United States as to taxes, that conservation can never be accomplished so long as a local government . . . must as a means of its financial survival get new tax base, new development, new payrolls into its boundaries.

Adding point to McCloskey's observation, it was revealed in *The New York Times* of October 14, 1969, that the administration was holding back essential information on environmental problems. A report on acid drainage from coal mines had been held back for nearly a month, apparently to cool down a rising demand in Congress for more money to fight water pollution.

A few weeks earlier the Department of the Interior had belatedly released another report on mine water pollution after its existence had been revealed by Ralph Nader.

This was the third incident. The previous spring a report came to light that had been kept under wraps for five months. The report had recommended that a major iron-producing company be enjoined from dumping ore waste into Lake Superior.

During the Disney-Mineral King litigation the Sierra Club had to threaten the Department of Agriculture with action under the Freedom of Information Act before Agriculture would relinquish some of its interdepartmental reports bearing on the case. Again, the Sierra Club had to pressure the Department of the Interior for information about the Santa Barbara oil leaks.

The Conservation Foundation has complained to the National Air Pollution Control Administration that the public's right to information was being throttled by hearings on air-quality control that were being held in semisecrecy.

Appalled by this demand for open hearings, government officials protested that they should be able to hold discussions with colleagues and write memoranda that might be considered confidential. At least one industry went further in public. Republic Steel complained at a hearing in Cleveland on the pollution of Lake Erie that charges of negligence in waste treatment were made public instead of being discussed privately.

There is of course a difference between a few memos or "working papers" passed around a department and an official report signed by a department head and prepared by people on the public payroll that is suppressed because it treads on important toes.

The new militancy on the part of supposedly harmless bird watchers has lit some fires in the breasts of their victims. Asked about the court actions of the Sierra Club, a spokesman for Standard Oil of California said, "We have no reactions that would be printable."

New York's public energy utility, Consolidated Edison, was equally bitter about the obstructions strewn in its way by conservationists in its efforts to get new sites for power plants. Con Ed is undoubtedly sincere in saying it is at its wit's end in the dilemma of supplying more power while being fought on plant sites. But the conservationists say that the real trouble at Con Ed is bad management, the solutions proffered are not realistic, and the utility has refused to explore much better alternatives offered to it.

Of course a fight brings out the militancy in both sides. The conservationist who wants to save a stream or a forest is not usually obstructionist, nor is he trying to drive a worthy industry out of business. Unfortunately, attitudes on both sides stiffen when the slugging starts.

But conservationists are more reasonable than they are given credit for being. They are not trying to turn back to the seventeenth century and disown all technology. Even nature's apostle Jean Jacques Rousseau didn't believe in that. As Jacques Barzun has said: "Rousseau never intended we should return to living in caves and wearing skins. This is neither possible nor desirable, but Rousseau saw that the complications of life resulting from civilization disturbs or destroys something valuable, something that cannot be flouted with impunity."

Faced with our dilemmas of shrinking space and a huge and growing population to feed, clothe, and house, abandoning technology is obviously not the solution. Our problem is not technology *per se* but the misuse of technology, what ecologist Kenneth Watt has called the ecocidal asymptote.

According to Watt, if you plot mathematically the exploitation of every natural resource, you can show it as two curves on a graph. One curve shows the rate of depletion, the other shows the technological capacity for exploiting it. Both curves start to rise gradually but then become steeper and steeper until they virtually "explode" into perpendicular ascent. Moreover, they follow the same curve pattern, maintaining parallel rates.

Translated, this means that the exploitation of a resource begins slowly, because the supply seems endless at first and the harvesting techniques are crude. As efficiency improves, the yield increases and depletion quickens with it. Dwindling supply spurs the invention of ever-more-efficient techniques, and their efficiency matches the depletion of the resource, until the peak is reached simultaneously for both curves with the extinction of the resource.

You can plot these curves for wildlife, for ocean fish, for timber, coal, oil, or any resource. The exhaustion of one resource aggravates the pressure on another. As is always true in ecology, every crisis is linked to all others. None can be solved alone.

Obviously, what is needed is a more responsible technology, one aware of the ecological consequences of exploitation. Watt believes that the energy needed just to keep our system going is now greater than the energy used in actual production. In other words, the overhead has become greater than the yield, and we are facing a disastrous "ecosystem depression."

It has become suddenly and frighteningly obvious that we are, after all, monkeys playing with a computer, entranced by the flashing lights and the action, but with very little understanding of what we are doing. It is also obvious now that we are all passengers on a crowded little globe whirling in space, a globe that is getting smaller by the hour. If we do not protect this, our only home, we will have no place to go.

We need, among other things, an ethic of the land. The air belongs to all of us, the sun belongs to all of us. What, then, of the land? We need, at the least, a land ethic, an acceptance of our responsibility for, and our need for, the land that makes our existence possible.

In *A Sand County Almanac*, Aldo Leopold says: "We abuse land because we regard it as a commodity belonging to us. When we see land as a *community* to which we belong, we may begin to use it with love and respect. There is no other way for land to survive the impact of mechanized man . . . that land is a *community* is the basic concept of ecology, but that land is to be loved and respected is an extension of ethics."

If ethics ennobles man, reason says it is not likely, in our lifetimes, to win any substantial victories over self-interest. Our morality sanctions exploitation and permits the violation of basic human rights by those powerful enough to bend the code in their direction. Much of the violation is committed by people convinced of the purity and righteousness of their motives and whose enormous acquisitiveness has so largely contributed to the murder of the environment.

The law is an extension of politics, and considering the ineffectual approaches of politics to our problem of ecocide, it is more than likely that in addition to a new way of looking at technology we need a new politics—a politics, if you will, of ecology.

There are signs of this happening, too, since the new generation is turning its radicalism away from Marx and Marcuse to Paul Ehrlich and Barry Commoner.

The politics of ecology go far beyond the relatively simple problems of pollution or open space, except as all problems are linked. Such politics would require consideration of every political act as it affects every aspect of our lives. Thus a proposed new highway would never be left to the engineers, to be considered solely in its role of usefulness to the motorist. It would have to be evaluated in all its consequences—its impact on the community through which it passes, its effect on traffic patterns and density, its smog and noise potentials, and the price society has to pay for its convenience to drivers.

If we can achieve these two things at least—a land ethic and a politics of ecology—we will have made one small step forward for mankind.

Index

Abalone, 135, 162–63
Abercrombie, Sir Patrick, 184
Acetic acid, 112
Acids, 50, 195
Actinomycetes, 104
Adams, John, 245
Adirondack Forest Preserve, 276
Adirondack Hudson River Association, 36–37
Adirondack Mountain Club, 38
Adirondacks, 14, 36–37
Advisory Committee on Poisonous Substances Used in Agricultural and Food Storage, 223
Africa, 15, 35, 103
  See also specific countries
African lion, 144
African realm, 15
Agnew, Spiro, 151
Airplanes, 23–24
  See also Jet planes
Air pollution
  in cities, 170–72
  from Everglades jetport, 80
  from Lower Manhattan Expressway, 199
Airports, dusting of, 231–32
Alabama, 245
Alaska, 43, 108, 143, 156–61, 223, 246, 252–53
  oil pipeline in, 296
Alaskan brown bear, 156
Alaskan seal, 160
Aldehydes from jet planes, 80
Aldrin, 215
Aleutian Islands National Wildlife Refuge, 162
Alewife, 20–21
Allagash (Maine), 45
Allen, Ivan, Jr., 168
Alligators, 72, 77, 145–46, 148–50, 280
Allin, Roger, 74, 76
Alum, 58
Aluminum, 96
American Automobile Association, 192
American Chemical Society, 60
American Forest Institute, 98
American Forest Products Institute, 98
American Forestry Association, 38, 118, 251, 284
American Machine and Foundry, 64
American Smelting and Refinery Company, 266–68
American Standard, 65
Ammonia, 50
Amory, Cleveland, 162
Amsterdam, 185–86
Anadromous fish, 138
Anderson, Clinton, 32
Anderson, Forrest H., 291
Andrews, James, 235
Anhinga, 67, 72
Anhinga Trail, 72
Anopheles mosquito, 225–26
Antarctic, 15
Antelope, 14, 143, 163
Anthony's Nose Mountain, 35

Antibiotics, broad-spectrum, 114
Ants, 105, 232–33
Apache, 270
Apes, 15
Appaloosa, 42
Applegate, Vernon C., 20
Aquifer, 47, 69, 76
Arabia, 63
Arcata Redwood Company, 124, 130
Arctic, 14, 158–59
Arizona, 28–34, 217, 270
Armadillos, 280
Arno River, 292
Arsenic, 215
  in drinking water, 62
Asgard, 103
Asheville, 172
Asia, 15
Aspen, 109
Aspinall, Wayne, 29, 34, 253–54
Aspirin, 114
Assateague Island, 260–61
Aswan High Dam, 10–12
Atkinson, Brooks, 38
Atlanta, 200
Audubon, John James, 140–42
Audubon Society, 38, 73, 76, 83–85, 118, 216, 281, 293
Augur buzzard, 154
Australia, 19, 144–45, 217
Australian realm, 15

Bacteria, 16–17, 56, 104–5
Badlands marsh (Illinois), 175–76
Bald cypress, 280
Bald eagle, 221, 223
Baljet, Peter J., 86
Baltic Sea, 292
Baltimore, 182, 196–97, 201–3
Barber pole worm, 78
Barbour, John, 163
Bark beetles, 237
Barnegat Bay, 56
Barnes, Will C., 155
Barron's, 223–25
Bartlett, E. L., 158
Barton, David, 202
Barzun, Jacques, 298
Basswood trees, 107
Batts, H. Lewis, Jr., 217
Bay Conservation and Development Commission, 178
Bayley, Ned D., 223
Bayous, 280
Bear Mountain, 35
Bear Mountain Park, 273
Bears, 43, 143–44, 156–59, 280, 286
Beaver, 27–28
Beaverhead River, 211
Beaverkill Creek, 207
Beech trees, 107, 141
Beechy Bottom, 273
Bees, 16, 232
Beetles, 19, 105, 227–28, 237
Behrens, E. F., 99
Benarides, Felipe, 145
"Benefit-cost" ratio, 203–7
Benzine, 215
Bering Strait, 15
Berkeley, 176–77
Berlin, Germany, 176
Berry, Phillip, 169, 296
Big Cypress and Everglades, 85–87
Big Hole River, 211
Big-horned bison, 143
Big Thicket, 279–82
Big Thicket Association, 282
Big Walnut Dam, 43
Binghamton, 44
Binhtri, 234
Biological amplification, 219
Biome
  definition of, 11, 13–14
  food supply in, 15–16
  habitats in, 106
  major zoologic areas, 15
Biosphere, divided into biomes, 13–15
Birch trees, 14, 106–7
Birds, 106
  jet planes and, 81
  scarcity of, 140–42
  succession of, 107
  See also specific birds
Biscayne aquifer, 69, 76, 85–86
Biscayne Bay, 295
Bison, 14, 18, 143–44
Bitumens, 249
Black bears, 156
Black-crowned night heron, 163
Black-footed ferret, 163
Blackfoot Rivers, 211

Black Forest, 111
Black Rock Forest, 37–38
Blacks, unemployment of, 168
Bloodworms, 51
Blue grouse, 291
Blue whale, 136–27
Bobolink, 107
Bodovitz, Joseph, 179
Body lice, 226–27, 237
Boeing Airplane Company, 174
Boeing 707 class plane, 23–24
Bolivia, 145
Boll weevil, 228
Bollworm, 238
Boone, Daniel, 40, 245
Borneo, 12, 103, 144
Boston, 182, 197
  cars in, 193
Bowers, William S., 237
Bowman, James H., 261
Bradley, Gen. Omar, 192
Bradley, Richard, 32
Branches of trees, 105
Brazil, 144
Breakneck Mountain, 35
Bridge Canyon Dam, 28–29, 31–34
Bridwell, Lowell J., 199
British Isles, *see* Great Britain
Broad-spectrum antibiotics, 114
Brooks, Paul, 99
Brooks Range, 223–24
Broun, Maurice, 154
Browder, Joe, 84–85
Brower, David, 34
Brown, Pat, 285
Brownlee Dam, 41
Brown pelicans, 155, 221
Brush control, 234
Buffalo River, 51
Buffalo Tiger, Chief, 82
Bull Creek, 131
Burns, chemical, from pesticides, 218
Burns, Hayden, 74–75
Business Week, 33
Butcher, Russell, 128, 131
Buzzard, augur, 154

Cabbage worms, 228
Cactus, 280
Cahalane, Victor H., 147, 156
Cahow, 163
Calcium metabolism, 221
Calhoun, John B., 180
California, 14, 28–29, 57, 63, 213–14
  birds in, 155
  pesticides and, 217
  redwood forests in, 14
    botany, 121–23
    destruction, 116–21, 123–31
    kinds, 121
California Aqueduct Enlargement, 28
California Redwood Association, 120
Calipee, 150
Cambridge, 200
Camels, 143
Cameron, Roderick, 295
Campbell, James, 42
Campbell Soup Company, 230
Camp Pendleton, 130
Canada, 15, 21, 138–39
  curbs detergents, 293
  pesticides and, 217
  polar bears and, 158
Canyon De Chelly, 173
Cape Cod, 260–62
Cape Cod Canal, 55
Cape Hatteras, 260
Caras, Roger, 151
Carbon dioxide, 24, 105
Carbon monoxide
  from cars, 170–71
  from jet planes, 80
Carcinomas, patients with, 218
Carey Act, 247
Caribou, 21–22, 43, 159
Carmer, Carl, 38
Carnac pine forest, 22
Carnivores, 16, 106
Carp, 55
Carr, John, 224
Carrion, 222
Cars, 185–86, 189
  banned in Yosemite, 258
  manufacturers of, sued, 295
  pollution from, 170–71
  See also Highways
Carson, Rachel, *Silent Spring*, 216, 223–24
Cassidy, Lt. Gen. William F., 76
Castle, Walter, 248
Castle Peak, 266, 269
Caterpillar, tent, 22–23

Catfish, 19
Cats, 15, 106
  See also specific cats
Catskill Aqueduct, 39
Catskills, 14
Catton, William R., Jr., 275–76
Cavanagh, J. P., 168
Caviar, 135
Cedar trees, 108
Centerville, Utah, 109–10
Central Arizona Water Project, 28–32
Central and Southern Florida Flood Control District (FCD), 70–71, 73, 75–76, 78, 85–86
Ceylon, 225
Chanute, Kansas, 63
Charlotte Dam, 43–44
Chase Manhattan Bank, 179
Cheetahs, 145–46
Chemical burns from pesticides, 218
Chemical defoliation, 233–35
Chemicals, *see specific chemicals*
Chemical Week, 120–21
Chester Beatty Research Institute, 222
Chestnut blight, 107
Chestnut-sided warbler, 107
Chestnut trees, 14
Chevron Oil Company, 295
Chicago, 189, 197, 201
Chickens, 229–31
China, 12, 104, 115
Chinook, 21, 55
Chloramphenicol, 114
Chlordane, 215, 220, 232
Chlorinated hydrocarbon pesticides, *see specific pesticides*
Chlorophyll, 15–16, 105, 222
Cholinesterase, 219
Christenson, Ray, 43–44
Chronic level of pesticides, 228
Church, Frank, 42
Cities
  Federal aid for, 167–68
  pollution of, 169–70
    air pollution, 170–72
    noise pollution, 172–74
    sprawl, 179–89
    waste disposal, 174–79
  See also Highways
Citizens for Clean Air, 172
Citizens' Committee for the Hudson Valley, 209
Civet cats, 280
Clams, 51, 219–20
Clancy, Peter J., 148
Clark, John R., 54–55
Clarkeson Engineering Company, 285
Clark Fork River, 211
Classen, Edward C., 145
Clausen, Don H., 118
Clear-cutting, 102–4
Clement, Roland C., 154, 239
Cleveland, 51, 189, 200
Cliff, Edward P., 264
Climax forests, 106–7
Close, Frederick J., 182
Cluster development, 183–84
Coal, 249
Coastal redwood (*Sequoia sempervirens*), 121
Cody, Buffalo Bill, 18
Cohelan, Jeffrey, 118
Cohen, Alfred B., 146
Cohesion principle, 122
Coho, 21, 219, 224
Cole, Lamont C., 12, 23–24
Collier County, Florida, 73, 78
Colorado, 29, 204–7, 246
Colorado Magazine, 264
Colorado Open Space Coordinating Council, 264
Colorado River, 28–35
Columbia, Maryland, 189
Columbia River, 55, 139
Commoner, Barry, 170, 300
Concorde (plane), 81
Condominiums, 184
Condor, 152–53, 286
Conifers, 14, 106
Connally, John, 282
Connecticut River, 51–52, 138
Conservation Foundation, 38, 296–97
Conservation and Research Foundation, 296
Consolidated Edison, 37–40, 52, 55–56, 208, 297–98
Control of nature, 12–13
Copepods, 16, 72
Copland, Aaron, 38
Corinth, New York, 53
Corkscrew Swamp Sanctuary, 73–74

Corn borer, 228, 237
Cornwall, 37–40
Costikyan, Edward N., 168
Côte d'Azur, 293
Cottam, Clarence, 148, 220, 231
Cougars, 22, 145–48, 286
Council on Environmental Quality, 61–62, 98
Courier-Journal, 40
Court action, 293–97
Courtney, Robert, 235
Coyotes, 22, 109, 159, 163
Crabs, 17, 135
Crafts, Edward C., 95
Craig, George B., Jr., 237
Crayfish, 55
Crickets, 105
Cromwell, Oliver, 184
Cross-Florida Barge Canal, 295
Crown Center project (Kansas City), 182
Crown Zellerbach Corporation, 95, 113
Crustacea, 16
Curry Company, 257–58
Cutting practices in forests, 101–3
Cuyahoga River, 51
Cyanide, 50
Cypress trees, 14, 73, 280

2,4-D, 235
D'Alesandro, Thomas, 203
Dams
  in Alaska, 43
  beaver, 27–28
  along Colorado River, 28–34
  earthquakes and, 34–35
  for flood control, 45–48
  in New York
    Charlotte Dam, 43–44
    along Hudson River, 36–40
  salmon and, 139–40
  along Snake River, 40–42
  See also specific dams
Dan Ryan Expressway, 197
Darwin, Charles, 9–10
Day, Ernest E., 266
DDE, 219, 225
DDT, 155, 215–27, 292
  used in Borneo, 12
  court suit on, 294–95
  in Montana, 291
Dechlorane, 233
Decibels, 172–73
Deciduous trees, 106
Deer, 106, 109, 143, 163
  in Everglades, 72, 77–78
  numbers of, 140–47
  white-tailed, 14
Defoliation, chemical, 233–35
Deichmann, William B., 218
Deinema, J. W., 287
Delaware River, 58
Del Norte Coast Redwoods State Park, 119
Dengue, 232
Denmark, 292
  pesticides and, 217
  salmon and, 138–39
Density, 179–81
Dermatitis from pesticides, 218
Desalinization, 63–65
Desecration of land, 243–44
Desert Land Act (1877), 247, 250, 252
Deserts, 14
Desmids, 16, 222
Detergent industry, pollution from, 50
Detergents, 293
Detroit, 189
  cars in, 193
  housing in, 167
Detroit River, 51
Diatoms, 16, 219, 222
22,25-Diazacholesterol, 237
Dickson, Felice, 79
Dieldrin, 135, 215, 217, 220, 226, 230–32
Digitalis, 114
Dimethyl sulfide, 113
Dimethyl sulfoxide (DMSO), 113
Dingell, John D., 97–99, 231
Dislocation by highways, 196–203
Disney, Walt, 283
DMSO (dimethyl sulfoxide), 113
Docking, Robert B., 51
Dodona, 103
Domination of man over nature, 12–13
Donora, Pennsylvania, 170
Dorr-Oliver, 65
Douglas, Marjory Stoneman, 68, 72
Douglas, William O., 40, 42, 48, 282
Douglas fir two-by-fours, 96

Douglas firs, 108, 111, 125, 127
Dow Chemical, 65
Doyle, Mortimer B., 271–72
Drainage and Barge Canal C-111, 75–76
Drug industry, 236
Drugs, 114
Drury, Newton B., 126
DuBridge, Lee A., 61, 233–34
Dunderberg Mountain, 35–36
Dunkle, Frank, 291–92
Du Pont, 65
Dust in New York City, 170
Dutch elm disease, 107, 237

Eagles, 144, 155, 159, 163, 221, 223
Eagles Nest Wilderness Committee, 264
Earnest, Russel, 228
Earthquakes and dams, 34–35
Earthworms, 104–5
Eastex, Inc., 282
East Meadow Creek, 264–65, 269
Ecosystem of forests, 104–12
Edge, Mrs. C. N., 154
Edgemere, New York, 175
Eel Dam, 43
Eel River, 127–28
Eggs and pesticides, 154–55, 221, 224, 229
Egrets, 67, 72, 76, 280
Egypt, 10–12
Ehrlich, Paul, 135, 300
Elephants, 15, 144
Elizabeth, Queen of England, 184
Elk, 129, 143, 147, 291
Elm trees, 14
Embarcadero Freeway, 200
Endangered species, 142–63
Endosulfan, 229
Endrin, 215
English sparrows, 19
Environmental Control Administration, 62
Environmental Defense Fund, 216–17, 294–96
Environmental Policy Act, 98
Environmental Protection Agency, 63
Environmental Quality Control Council, 61, 288
Eremotherium, 143
Erickson, Eldon L., 268–69
Erickson, Ray C., 142
Eskimo curlew, 163
Eskimos and polar bears, 158
Estrogen, 221
Ethiopian realm, 15
Everglades, 88, 262
  Big Cypress and, 85–87
  description of, 66–68
  hunting in, 71–78
  jetport in, 78–84
  Lake Okeechobee and, 68–78
    water shortage, 86–87
Everglades Kite, 221
Evergreens, 14
Ever-Normal Trough, 195
Evolution, 91
Explorers, Spanish, 68

FACE (Facts, Action, Communication and Evaluation), 98
Falcons, 221, 223–24, 229
Farb, Peter, 70, 108, 144
Farmington, Utah, 109–10
Farsta, 187
Fat and DDT, 219
Fault areas and dams, 34–35
FCD (Central and Southern Florida Flood Control District), 70–71, 73, 75–76, 78, 85–86
Federal aid to cities, 167–68
Federal Aviation Administration, 78
Federal Aviation Authority, 173
Federal Power Commission, 38–42
Federal Property Review Board, 251
Federal-state joint action, 45
Federal Trade Commission, 97
Federal Water Pollution Control Administration, 61
Ferrets, 163
Fertilizer plants, 172
Fertilizers and Everglades jetport, 80
Filter feeders, 220
Finch, Robert H., 215
Fingernail clams, 51
Finland, 292–93
Fin whales, 137
Fire ant, 232–33
Fire Island, 260
Fires, forest, 108
  redwoods and, 122, 126–27

Fir trees, 14, 108–9, 111, 125, 127
Fish, 16
  in Everglades, 72
  killed by thermal pollution, 55–56
  See also specific fish
Fish-eating birds, 155
Fisk University, 197
Fitzgerald, William, 146–47
Flash distillation process, 64
Flood control
  dams and, 45–48
  in Everglades, 69–78
  forests and, 109–10
  redwoods and, 126–28
Flood plains, 46–48
Florence, Italy, 45
Florida, 19
  alligators in, 149–50
  pesticides and, 217
  See also Everglades
Florida Key deer, 163
Florida Power and Light Company, 295
Flowers, 16
Floyd Bennett Field, 175
Fluoride pollution, 172
Flying squirrels, 280
Fog and redwoods, 116
Folkestad, James O., 264
Food and Agricultural Organization, 222–23
Food chains, 16
Food pyramid, 15–16
Food supply, 15–16
Forbush, Edward Howe, 140
Ford, Henry, II, 193
Ford Foundation, 126
Ford Motor Company, 171
Forest fires, 108
  redwoods and, 122, 126–27
Forest reserves, 250
Forests, 115
  cutting practices in, 101–3, 251
  ecosystem of, 104–12
  fringe products of, 112–14
  multiple use of, 92–93, 95–100, 264–65
  mythology about, 103–4
  National Forest Timber Supply Act and, 91–101
  pigeons and, 141
Forests (*cont.*)
  of redwoods, 14
    botany, 121–23
    kinds, 121
    destruction, 116–21, 123–31
  in Norway, 111–12
  in southern U.S., 78–79, 88
  in Sweden, 293
Forked River, New Jersey, 56
Fort Lauderdale, 69
Fort Ord, 130
Foster City, California, 179
Fourteenth Amendment, 294
Foxes, 22, 106, 143
Fraley, Pierre, 111
France, 138, 217, 229, 245, 293
Frankfurt, Germany, 185
Franklin, Benjamin, 245
Freezing and desalinization, 64
Fresh Kills, New York, 175
Friends of the Earth, 296
Friends of the Sea Otter, 162
Frogs, 106
Frome, Mike, 270, 289
Fruit fly, Mediterranean, 19
Fund for Animals, 162
Fungi, 16–17, 104–5
Fur industry, 145–47, 160

Gadsden Purchase (1853), 246
Garelik, Sanford D., 176
Garfish, 72
Gas, 249
Gasoline, 171, 296
Gateway National Recreation Area, 175
General Electric Company, 174
General Motors Corporation, 171
General Sherman (redwood tree), 121
George, King of England, 244–45
Georgia, 196
Georgia-Pacific Corporation, 95, 97, 124–26, 130–31
Germany, 292
Gilliam, Harold, 176
Ginkgo tree, 104
Glasgow, 186
Glen Canyon Dam, 30–32
Gold Bluffs, 129
Golden marmoset, 144
"Gooley No. 1 Dam," 36–37

Gordon, Clarence C., 240
Gore Range, 264
Gorillas, 146
Grand Canyon, 29, 31–32, 35, 271
Grand Coulee Dam, 139–40
Grasslands, 14
Grassy plants, 104
Gray squirrel, 14
Grazing rights and multiple use of land, 270–71
Great Britain, 111, 138–39, 245–46
Great Kills, New York, 174–75
Great Lakes, 19–20, 51, 224
 See also specific lakes
Great Smoky Mountain National Park, 208
Great Swamp, New Jersey, 278–79
Greece, 35
Green Belt, 184–85
Green Belt Act (1938), 184
Green turtles, 135, 150–51
Grizzly bear, 156–57, 286
Gross national product, 195
Ground squirrel, 14
Grouse, 107, 291
Gruening, Ernest, 43
Gulf American Land Company, 73
Gulick, Mrs. Charles, 177
Gulls, 219, 229
Gum Slough, 84
Gunter, Wilbur L., 232
Gutermuth, C. R., 100
Gypsy moth, 107–8, 225–26, 238

Habitats, 106
Hamburg, Germany, 185
Hano, Arnold, 288
Hansen, John, 257–58
Hansen, Orval J., 267
Hansen, Roger P., 265
Hardin, Clifford M., 93–94, 98–99, 215, 264, 287
Hardwoods, 106–7, 111, 281
Harp seal, 159–61
Harris, Robert C., 87
Harrison, William, 250
Harry, Bryan, 258
Hartzog, George, Jr., 77, 256, 262
Hawaiian dark-rumped petrel, 145
Hawk Mountain Sanctuary, 154
Hawks, 22, 106, 153–55, 221–22
Hayes, Weyland J., 224
Headgate Rock Dam, 29
Hell's Canyon, 40–42
Hell's Canyon Preservation Council, 42
Helsinki, 185
Hemlock trees, 14, 106–8, 112, 141
Hendee, John C., 275–76
Henderson, Richard, 245
Hendersonville, North Carolina, 171–72
Heptachlor, 215, 226, 233
Heptachlor epoxide, 230
Herbivorous animals, 15–16
Herbs, 104, 109
Herley, Dempsie, 282
Hermit crab, 17
Herons, 67, 72, 163, 229, 280
Herring, 16
Herring gulls, 219
Hessian Lake, 273
Hetch Hetchy Reservoir, 271
Hexachloride, 215
Heyerdahl, Thor, 136
Hickel, Walter J., 42, 61–62, 77, 81, 83, 146, 208, 255, 287–88, 296
Hickory trees, 14
Hicks, Robert B., 285–86
High-blood-pressure patients, 218
High Mountain Sheep site, 41–42
High Tor, 36
Highways, 191–214
 as big business, 194–96
 dislocation by, 196–203
 need for, 192–94
 through open country, 203–14
Highway Trust Fund, 195
Hill, Gladwin, 62, 253
Hill View Reservoir, New York, 59
Hippopotamus, 15, 145
Hodges, Ralph, 97
Hoff, Phillip, 296
Hoffer, Eric, 12–13
Holland, 137–38, 293
Holmes, J. H., 218
Homestead Act (1862), 246–47, 252
Homesteaders, 250
Honey bees, 232
Hood seals, 161
Hoover Dam, 29–30, 33–35
Horses, 143, 155–56
Hosmer, Craig, 33

Housefly, 226
Housing
building abandonment, 167
lumber shortage and, 93–94, 101
urban sprawl and, 179–89
Houston, Sam, 280
Howell, John H., 20
Hualapai, 34
Hudson River, 35–40, 51–56, 58–59, 208–10
Hudson River Tourway, 59
Hudson River Valley Commission, 36
Hudson Valley, 59
Humanization, 12
Humboldt State Park, 129, 131
Humidity, 105
Humpbacked whales, 137
Humus, 105, 108
Hunter, Marjorie, 253
Hunting, 18
in Everglades, 77–78
Hurricane Donna, 70
Huxley, Thomas, 10
Hydra, 17
Hydrocarbons, 170–71
from jet planes, 80
Hydroelectric power
air pollution from, 170, 195
Con Ed and, 37–40, 52
feasibility of, 32–33
price of, 32
from Ramparts Dam, 43
Hydrogen, 105

Ibis, 72
Imperial Dam, 29
India, 12, 35, 104, 225
Indiana, 228
Indian elephant, 15
Indian Point, New York, 55
Indians, 270
Everglades jetport and, 81
forests and, 104
land bought from, 278–79
land rights of, 247–48
Indian tiger, 15, 144
Industry
air pollution from, 170
water pollution from, 50
See also specific industries and companies
Insecticides, *see* Pesticides
Interdependence of living creatures, 16–24
International Commission for Northwest Atlantic Fisheries, 138
International Union for the Conservation of Nature, 140
International Union for Conservation of Nature and Natural Resources, 147
International Whaling Commission, 137
Interstate Highway 40, 197
Interstate Highway 70, 204–7
Interstate Highway 93, 207–8
Interstate Highway 95, 196
Interstate Highway Program, 194–96
Intestines of man, 17
Iowa, 59
Iran, 225
Iraq, 225
Ireland, 138
Irradiation, 236
Isle Royal National Park, 109
Israel, 11, 63, 155
Italy, 45, 217, 292–93
Ivory-billed woodpecker, 281

Jack pine, 108
Jackson, George, 156
Jackson, Henry M., 120, 255
Jaguars, 145–46, 280
Jamaica Bay, New York, 174–75
James, Jim, 285
James, M. R., 287
Japan
DDT and, 217
fish and, 135–36, 293
logs exported to, 95
whales and, 137
Japanese beetle, 19
Jarfalla, 188–89
Jay, John, 245
Jedediah Smith Redwoods State Park, 119, 129
Jefferson, Thomas, 245–46
Jersey Central Power and Light Company, 56
Jet planes, 23–24
air pollution from, 80
noise pollution from, 81, 173

Jetport in Everglades, 78–84
John J. Fitzgerald Expressway, 196–97
John Pennekamp Coral Reef, 71
Johnson, Lyndon B., 44, 56, 61, 117–19, 261
Johnston, Velma, 156
Jones Falls Expressway, 197
Jordan, Len, 42
Jordan, Robert, III, 77
Judy, Richard H., 82, 84
Juriens, Switzerland, 110–11
Juvenile hormone, 236–37

Kaibab forest, 109
Kaibab Industries, 264
Kanab Diversion Project, 31
Kangaroo, 15, 144–45
Kansas, 51, 59, 246
Kariba Dam, 35
Kearing, Samuel J., 174–75
Kelly Air Force Base, 231
Kenai National Moose Range, 250
Kennedy, John F., 119, 227
Kennedy, Robert, 119
Kennedy Airport, New York, 175
Kensico Reservoir, New York, 58
Kent, William, 271
Kentucky, 245
Kenya, 154
Kerr, Mrs. Clark, 177
Kimball, Thomas L., 127
Kimble, Glenn, 151
Kinglets, 107
Kingston, New York, 52
Kirby Lumber Company, 281–82
Kirk, Claude R., Jr., 77, 82–83
Kissimmee River, 85
Kiwi, 15
Klamath County, California, 101–2
Klamath River, 127–28
Klein, Karl L., 61
Knauer, Virginia, 62
Koala bear, 15, 144
Koch, Edward I., 199
Koyna Dam, 35
Kratz, Lawrence, 268
Kreig, Margaret, 114
Krill, 137
Kuchel, Thomas H., 285
Kuperberg, Joel, 73

La Jolla, 171
Lake Baikal, 293
Lake Erie, 19, 51, 57, 229
Lake George, 229
Lake Havasu, 31–33
Lake Huron, 20
Lake Kremasta, 35
Lake Mead, 30–31, 33–35
Lake Michigan, 20–21
  DDT in, 219, 224
Lake Mojave, 31, 33
Lake Okeechobee, 67–78, 86–87
Lake Ontario, 19
Lake Passaic, 278
Lake Powell, 30–31
Lake Tahoe, 51
Lake Tear of the Clouds, 35
Lamprey, 19–20
Land and Water Conservation Fund Act (1968), 261
Land grants, 245
Land Ordinance (1785), 246
Larva, 236–37
Las Vegas, 63
Laurens, Henry, 245
Law, John H., 237
Lawrence, John W., 199
Lead, 170–71
Leaves, 105
Lebanon, 11
Leguminous plants, 16–17
Leopard, 145–46
Leopold, Aldo, 13, 299
Leukemia patients, 218
Levittown, Long Island, 182
Lewis, Gerald, 150
Lewis, Richard, 235
Lice, 17, 226–27, 237
Life, 120
Lightsey, Harry M., 232
Lignin, 113, 237
Lindane, 215, 220
Lindau, Jules W., IV, 232
Lindsay, John, 57, 145, 168, 191, 198, 227
Lion, African, 144
Little Boulder Creek, 267
Litton, Martin, 290
"Livable Park and Seashore Plan," 118–19

Liver malfunctions and DDT, 218, 221, 225
Lizards, 106
Lobbying, highway, 194
Loire Valley, France, 229
London, 138, 171, 184–85
Long Island, 14
  roads on, 191
Los Angeles, 189
  black unemployment in, 168
  cars in, 193
Los Angeles County, 63
Louisiana Purchase (1803), 246
Lower Manhattan Expressway, 198–99
Low Hell's Canyon Dam, 41
Low Mountain Sheep Dam, 41
Lynch, Kevin, 181
Lynx, 163, 280

McAteer, J. Eugene, 178
McCloskey, Michael, 99, 286, 296
McCloskey, Paul, Jr., 296
Macek, Kenneth J., 228
McHarg, Ian, 46–47
McKenzie, Leonard, 260–61
McKinley, William, 250
McKusick, Robert, 235
McLaughlin, Mrs. Donald, 177
McLaughlin, John S., 287
Madison River, 211
Mad River, 127
Magnolia trees, 280
Magruder, William M., 173
Maine, 107, 138
Malaria, 225–26
Malathion, 232
Mallards, 225
Malone, James, 82
Mangroves, salt, 72
Maple syrup, 112
Maple trees, 14, 107
Marble Gorge Dam, 28–38
Marcuse, Herbert, 300
Margays, 146
Marine animals
  DDT in, 219–20
  scarcity of, 135–40
  See also specific marine animals
Marshes, 47
  See also specific marshes
Martin, Paul S., 142
Marx, Karl, 300
Maryland, 217
Mason, John A., 232
Mason bill (New York), 145–46
Matthiessen, Peter, 85
Mayors, 168
Meat-packing plants, pollution from, 59–60
Mediterranean, fishing in, 10–11
Mediterranean fruit fly, 19
Meharry Medical College, 197
Menhaden, 55
Merced River, 257
Mercury, 215, 291–93, 295
Mesquite, 280
Metcalf, D. R., 218
Metcalf, Lee, 118–19, 211
Methanol, 112
Methoxychlor, 220
Methyl acetone, 112
Meuse Valley, Belgium, 170
Mexico, 29, 31, 33, 136, 225, 246
Miami, 69, 78, 86
Miccosukee Indians, 81
Mice, 105
Michigan, 51, 217
Microencapsulated insecticide, 233
Midsummer Day, 104
Midwest, water shortage and pollution in, 50–60
Milan, Italy, 293
Mildew, 17
Milk, 220
Mill Creek Basin, 119, 129
Millipede, 105
Mineral King, California, 282–88
Mineral Leasing Act (1920), 249
Miners, 248–49
Mining industry, 251, 254
  multiple use and, 266–70
Mining Law (1872), 249, 252, 254, 267, 269–70
Mink, 280
Miscarriages, 235
Mississippi, 46, 224, 245
Mississippi River, 52, 55, 245
Missouri, 59
Missouri River, 59–60, 211
Mistletoe, 17
Model-cities program, 167

Molasses, 113
Moles, 105
Mollusks, 16
Molybdenum, 266–68
Monarch Creek, 286
Monkeys, 15
Monoculture, 227, 236
Montana, 210–12, 291–92
Monterey pine trees, 126
Moose, 43, 109, 143
Moritz, Ned, 231
Moses, Robert, 175, 198–99, 201
Mosquitoes, 225–27, 232, 237–38
Moths, 107–8, 225–26, 238
Motorcycle, noise pollution from, 173
Mounsey, William B., 264
Mountain lion, 147
Mountains, 14
  See also specific mountains
Mount Trashmore, Illinois, 175–76
Mourning doves, 140
Mud snail, 270
Mueller, Paul, 227
Muir, John, 9, 271
Multiple Use Act, 252
Multiple use of public lands, 250, 253–254, 263
  forests and, 92–93, 95–100, 264–265
  grazing rights and, 270–71
  mining and, 266–70
  parks, wilderness, and, 273–90
  single use and, 272–73
Multiple Use-Sustained Yield Act (1960), 264
Murphy, George, 285
Murphy, Thomas F., 209
Muskie, Edmund S., 98–99
Musk oxen, 158–59
Mutualism, 16–17
Mythology about forests, 103–4

Nader, Michael, *The Living Wilderness*, 272
Nader, Ralph, 296–97
Naftalin, Arthur, 168
Nashville, Tennessee, 197
Nashville warbler, 107
National Airport, 231
National Forest Conservation and Management Act, 93
National Forest Timber Supply Act, 91–107
National Parks Association, 83, 118, 284
National Reclamation Act (1920), 247
National Rifle Association, 93
National System of Interstate and Defense Highways, 194
National Timber Industry, 98
National Wilderness Act, 252
National Wilderness Preservation System, 252
National Wildlife Federation, 93, 127
National Wildlife Refuge system, 250
Nature Conservancy, 38
Nearctic realm, 15
Nebraska, 59
Nelson, Gaylord, 83
Nelson, Samuel B., 30
Nematodes, 104
Neotropical realm, 15
Neuzil, Dennis R., 204–7
Nevada, 29, 251
Newcomb, 36
New Hampshire, 207–8, 217
New Jersey, 14, 53
  roads in, 191
New Mexico, 29, 246, 270
New Orleans, 199–200
"New towns," 187–89
New York, 19, 36–37, 51–53, 57, 189
  deer in, 140
  pigeons in, 141
  sewage disposal in, 174–75
New York City, 36–37, 40, 172
  cars and, 191
  density of, 179–80
  dust in, 170
  housing in, 167
  open spaces in, 181
  temperature inversion in, 170–71
  urban renewal in, 183
  water shortage and pollution in, 57–59
Nez Perce site, 41–42
Niagara Falls, 51
Nile River, 10–11
Ninth Amendment, 294
Nitrogen oxides, 170–71
  from jet planes, 80

Nixon, Richard, 61, 83, 100, 167, 169, 251, 255, 261
Noise Pollution
  in cities, 172–74
  from Everglades jetport, 80–81
Nonpoisonous pesticides, 236–40
Nordwall, David, 264
Normann, A. C., 139
Norns, 103
Norse mythology, 103
North America, original animals of, 142–43
North Carolina, 140
Northwest Ordinance (1787), 245
Northwest Territories, 245
Norton, Boyd, 42, 268
Norway, 104, 111–12, 135
  salmon and, 138
  whales and, 137
Norway rat, 19
Nuclear plants
  for desalinization, 64
  for power, 32, 54–56
  thermal pollution from, 54–56, 295

Oakland airport, 177
Oakley Dam, 43
Oak trees, 107, 127, 141
Oak wilt, 107
Oceans, pollution of, 56–57, 136
Ocelots, 145–46, 280
Odemar, Melvin, 162
Ohio, 46, 51
Ohio River, 52
Oil, 50, 249
Oil companies, 171
Oklahoma, 246
Oklawaha River, 296
Olentangy River Valley, 200
Olmsted, Frederick Law, 197
Omnivores, 106
O-o-a-a (Hawaiian bird), 145
Open space, 181–89
Operation Whooping Crane, 152
Opossums, 140, 280
Orange trees, 172
Orangutan, 144
Oregon Compromise (1846), 246
Organophosphate plants, workers in, 218–19
Oriental realm, 15
Osmosis, reverse, 64–65
Osprey, 68, 155, 163, 221
Otters, 72, 280
Ottinger, Richard, 40, 56, 209
Outdoors Unlimited, 98
Owens, Philip, 73
Owings, Margaret, 162
Owings, Nathaniel, 202
Owls, 106
Ownership of land, 244–45
Oxbow Dam, 41
Oxygen
  in photosynthesis, 105
  reduction of, 222
  in streams, 50–51
Oysters, 135, 219–20
Ozarks, 14
Ozone, 170

Pacific Northwest Power Company, 41–42
Pacific Southwest Water Plan, 28
Padre Island, 260
Pakistan, 225
Palisades, 36
Palisades Interstate Park, 39
Palm Beach, 69
Palmetto palms, 280
Palo Verde, 29
Panther, 147
Paper mills, pollution from, 50, 53–54, 172, 195
Parasites, 17–18
Paris, 185
Parker Davis Dam, 29
Parks
  land for, 254–62
  multiple use and, 273–90
  See also specific parks
Passenger pigeons, 140–42
Patuxent Wildlife Research Center, 142
Pearl, Milton A., 253–54
Peekskill, New York, 52
Pelican Island Refuge, 250
Pelicans, 68, 76, 155, 221
Penguins, 14–15, 220
Penn, William, 278–79
Pennekamp, John, 71, 76, 86
Pennsylvania, 51
Peregrine falcon, 221, 223–24, 229

Peru, 135–36, 145
Peshtigo fire, 108
Pesticides, 50, 55, 106, 295
  birds and, 154–55
  chemical defoliation, 233–35
  DDT, 215–27
  in drinking water, 62
  Everglades jetport and, 80
  need for, 227–33
  nonpoisonous, 236–40
  See also specific pesticides
Peter Baran and Sons, Inc., 148
Peters, John C., 211–12
Peterson, Roger Tory, 107
Petris, Nicholas, 178
Pharmaceutical industry, 236
Pheasants, 140, 224, 291
Phenol, 50
Philadelphia, 182, 189, 200
  housing in, 167
Philippine eagle, 144
Philippines, 104
Phoenix, 29, 32
Phosphates, 50, 249
Photosynthesis, 16, 105, 222
Phthalic anhydride, 113
Phytoplankton, 222
Pickling liquors, 195
Pike, 51
Pimlott, Douglas H., 161
Pinchot, Gifford, 271
Pine sawfly, 238
Pine trees, 14, 69, 126, 141, 171–72, 281
Pittsburgh, 189
  cars in, 193
Plains, western, 14
Plankton, 16, 219
  increase in, 10–11, 56
Plastics, 96
Platypus, 15
Pleasant Valley Dam, 41
Pleistocene period, 143
Plywood, softwood, 94
Plywood sheathing, 96
Poage, W. R., 94, 97
Point Reyes, 261–62
Poisoning from pesticides, 218
Poland, 292
Polar bears, 14, 145–46, 156–58
Ponderosa pine trees, 171
Population explosion, 196, 293
Porpoises, 137, 161
Portola, Don Gaspar de, 123
Portugal, 138
Potassium, 249
Potato beetle, 227–28
Potomac National River Bill, 44
Potomac River, 44–45
Pough, Richard H., 154
Poultry, 229
Prairie Creek Redwoods State Park, 119, 128
Prairie dog, 14, 22, 163
Pratt, Jerome, 150
Predators, natural, necessity for, 18, 21–22, 106, 162–63
Prey, birds of, 153–54
Pribilof Islands, 159
Pribilof seal, 160–61
Price, Donald E., 264
Prickly Creek, 211
Prince Albert National Park, 22–23
Private utilities, 41
Propionic acid, 112
Public Land Review Commission, 253
Puma, 106, 109, 147
Pusey, Nathan M., 38

Quail, 224
Quercetin, 113
Quetzal, 163
Quinine, 114

Rabbits, 19, 72, 106
Raccoons, 72, 140, 280
Raftery, John C., 77, 79
Railroads, 247, 250
Rainbow Bridge, 30
Ramparts Dam, 43
Ranching on public lands, 248, 251
Rat, Norway, 19
Rawls, Oscar, 76
Reagan, Ronald, 119–20, 130, 178–79, 285
Reclaiming water, 63–65
Recycling wood products, 114
Redd, Esther, 202
Red Data Book, 147
Red River Gorge, 40, 43
Red squirrel, 14
Redstarts, 107

Redwood Creek Basin, 119, 129
Redwood forests, 14
  botany of, 121–23
  destruction of, 116–21, 123–31
  kinds of, 121
Redwood National Park, 117–18, 130
Reed, Nathaniel, 82
Reidel, Carl, 23
Reille, Antoine, 229
Reproduction, 136
  DDT and, 228–29
  See also Eggs; Pesticides
Republic Steel, 297
Reserpine, 114
Resor, Stanley R., 76
Reston, 189
Reuss, Henry S., 145, 174
Reverse osmosis, 64–65
Rhine River, 51, 55, 229, 292
Rhinoceros, 15, 144
Ribicoff, Abraham, 52
Ring-necked pheasants, 140
Ripley, S. Dillon, 152–53
Risebrough, Robert, 155
Rivers, 35
  See also specific rivers
Riverside Park, New York, 53
Riviera, 293
Roads, 191
  as big business, 194–96
  dislocation by, 196–203
  need for, 192–94
  through open country, 203–14
Roberts, E. F., 294
Robertson, James, 245
Robinson, Gordon, 107
Roche, James M., 193
Rockefeller, David, 198
Rockefeller, Laurance, 274
Rockefeller, Nelson, 53, 59, 145, 198, 209, 217
Rockland County, New York, 52–53
Rodents, 222
Rogers, Archibald, 202
Röller, Herbert, 237
Romney, George, 94, 99
Roosevelt, Theodore, 109, 147, 250
Roosevelt elk, 129
Roseate spoonbill, 68, 76, 280
Rosier, Lance, 282
Rothschild, L. D., 38
Rotifers, 219
Rousseau, Jean Jacques, 9, 298
Rowse, Arthur E., 230
Roxburgh, Duke of, 137
Ruffed grouse, 107
Ruhr River, 60–61, 293
Russia, *see* U.S.S.R.
Russian thistle, 19

St. Louis, housing in, 167
St. Petersburg, 78–79
St. Regis, 95
St. Regis River, 211
St. Thomas, Virgin Islands, 64
Saline soil, 12
Salinization of Colorado River, 32–33
Salmon, 20–21, 41, 43, 51, 135, 137–40, 159
  See also specific kinds of salmon
Salmon River, 41, 167
Salt mangroves, 72
Salt River, Arizona, 29
Salt River Valley, Arizona, 294
Salt water in Everglades, 69, 72, 75–76, 86
Samuelson, Don, 266–67
San Bernardino Mountains, 171
San Bruno Mountains, 179
Sandburg, Carl, 271
San Francisco, 176–77, 200–201
San Francisco airport, 177
San Francisco Bay, 176–79
San Mateo County, California, 178
Santa Barbara Channel, 161
Santa Barbara oil leaks, 297
Sausalito, 179
Savannah sparrow, 107
Save-the-Redwoods League, 118–19, 126, 130–31
Save San Francisco Bay Association, 178
"Saw grass," 67, 72
Saylor, John P., 99–100
Scalp Act (Pennsylvania), 153
Scenic Hudson Preservation Conference, 38–40
Schaefer, William D., 202
Schistosomiasis, 225
Schweitzer, Albert, 13
Scollay Square (Boston), 182
Scotland, 111, 138

Scott Paper Company, 95
Screwworm flies, 236
Scrub oak trees, 14
Sea anemone, 17
Seaborn-Wells project, 286–87
Seafood, scarcity of, 135–40
Sea lamprey, 19–20
Sea lions, 161–62
Seals, 159–61
Sea otters, 159, 162–63
Sea snails, 72
Seattle, 174, 182
Sea turtle, 150–51
Sediment, effects of, 11
Sequoia gigantea (Sierra redwood), 121
Sequoia National Park, 282–83, 287–288
Sequoia sempervirens (coastal redwood), 121
Sespe Creek, 153
Seton, Ernest Thompson, 141
Sevier, John, 245
Sewage
  amount per second of, 170
  dams and, 44
  disposal of, in cities, 174–79
  from Everglades jetport, 80
  killing streams, 50–51
  at parks, 257
  reclaiming water from, 63
Shark River Slough, 74
Shellfish, 72, 293
Shepherd, Donald R., 231
Sheridan, Gen. Phil, 18
Shorelines, 260
Short-nosed sturgeon, 145
Shrews, 105
Shrimp, 72, 135, 219
Shrubs, 106
Siberia, 143
Sierra Club, 31–33, 38, 42, 80, 83, 93, 117–19, 123–25, 128–30, 209, 216, 264, 281, 283–85, 287, 293–94, 297
Sierra redwood (*Sequoia gigantea*), 121
Silvex, 235
Silvichemicals, 112
Simpson (lumber company), 95
Singapore, 293
Single use and multiple use, 272–73
Sing Sing prison, 53
Sioux, 270
Sitka spruce, 112
Sive, David, 38, 209–10
Skärholmen, 188
Skuld, 103
Sláma, Karel, 238
Slopes, 47
Sloths, 15
Small Tract Act, 252
Smiley, Nixon, 71, 74
Smith, Anthony Wayne, 78, 119, 289
Smith, Bailey, 173
Smith, Hoke, 248
Smith, Spencer, Jr., 100
Smith River, 128
Snake River, 40–42, 139
Snakes, 106
Snow leopards, 146
Snowy egret, 280
Sodium, 249
Softwood lumber, 94, 111
Softwood plywood, 94
Soil, saline, 12
Soil bacteria, 16–17
Soil mites, 104
Sonic booms, 173
Sonoran pronghorn antelopes, 163
South America, 15
  extermination of animals in, 144
South Carolina, 196
South Florida Coordinating Council of Sportsmen's Clubs, 78
South Platte River, 60
South Swedish Forest Owners' Association, 112
South Vietnam, 233–34
Soviet Union, *see* U.S.S.R.
Spanish explorers, 68
Sparrows, 19, 107
Spear, as primitive weapon, 143
Sperm whales, 137
Sprawl of cities, 179–89, 243
Spring Creek, New York, 175
Springtails, 105
Spruce trees, 14, 109, 112, 125
Squirrels, 14, 72, 106, 280
SST, 81, 173–74, 296
Standard Oil Corporation, 297
Stanley, G. W., 281

Starlings, 19
State-Federal joint action, 45
Staten Island, 174–75
Steam plants, 32, 34
Stebbins, Robert, 213–14
Steel mills, pollution from, 50, 60–61, 195
Steelworkers, 172
Stewart, Alan C., 79
Stickel, Lucille F., 225
Stockholm, 185–88
Stockmen, 251
Stock Raising Homestead Act (1916), 247
Stomach disturbances and DDT, 218
Stone, Edward C., 126–27
Storm King Mountain, 35, 37–40, 52, 208
Stream Preservation Law (Montana), 212
Streptomycin, 114
Striped bass, 55
Sturgeon, 41, 145
Sublethal level of pesticides, 228
Subways
 in Europe, 185
 noise pollution from, 173
Suez Canal, 11–12
Suffolk County, New York, 213, 217
Suffolk County Mosquito Control Commission, 214
Sugar, 105
Sugar-beet industry, 60
Sulphur, 249
Sulphur dioxide, 172, 195
Sulphuric acid, 293
Sulphur oxide, 170, 172
Sumatra, 104, 144
Sun heat and desalinization, 63–64
Sun River Dam, 43
Superhighways, 193–94
Sweden, 112, 217, 227, 292–93
Sweet gum trees, 280
Symbiosis, 17
Syria, 11
Syringaldehyde, 237

2,4,5-T, 233–35
Tamiami Trail, 70
Tampa, Florida, 78
Tanhiep, Vietnam, 234
Tapeworm, 17
Tapiola, 189
Tapir, 143
Tarrytown, 209
Taylor Grazing Act (1934), 247, 249
Taylor Slough, 74
Technology and pollution, 170
Tellico Dam, 43
Temperature inversion, 170–71
Tener, John S., 158
Tennessee, 245
Tennessee Valley Authority, 172
Tent caterpillar, 22–23
Termites, 17, 105
Tetracyclines, 114
Texas, 63, 246
Texas blind salamander, 145
TFM, 20
Thermal pollution, 54–56, 295
Thor, 103
Thoreau, Henry David, 271
Thrushes, 107
Tiber River, 292
Tigers, 15, 144–46
Tilghman, Billy, 18
Timber Culture Act (1873), 249–50
Timber industry
 Big Thicket and, 281–82
 Federal government and, 254
 See also Forests
Timber and Stone Law, 250
Timor, 103
Titanium Lead Company, 56
Titanotylopus, 143
Toffler, Alvin, *Future Shock*, 196
Tokyo, 293
Tonto National Forest, 235
Tortugas shrimp, 72
Towell, William E., 95, 284
Towhees, 107
Toxaphene, 220
Trachoma, 225
Train, Russell E., 62, 261
Trans-Alaska oil pipeline, 296
Treasure Island, California, 177
Trees
 rise of water in, 121–22
 See also Forests; *specific trees*
Tree squirrels, 14
Trinity River, 127
Triparanol, 237

Trout, 55, 229
Troy, New York, 52
Truman, Harry, 69
Tucson, 29, 32
Tuna, 135
Tupelo tree, 280
Tupling, Lloyd, 98
Turkeys, 78, 229–30
Turner, Francis C., 207
Turpentine, 112
Turtles, 72, 135, 150–51
Tussock moth, 238
Twigs, 105
Typhus, 226–27

Udall, Stewart, 30, 33–34, 41, 43–44, 119–20, 152, 176, 209, 255, 272–73, 277, 283–84, 289
Ultrafiltration, 64
Unemployment of blacks, 168
United States Steel, 295
Unwin, Sir Raymond, 184
Uppsala, Sweden, 104
Urban renewal, 167, 182–83
Urban sprawl, 179–89
Urda, 103
U.S.S.R., 15, 135, 138, 292, 293
  polar bears and, 158
  whales and, 137
Utah, 29, 270

Vail, Colorado, 264
Valborgsmassoafton, 104
Vällingby, 187–88
Van Arsdale, Harry, 198
Van Duzen River, 127, 131
Vanillin, 112–13, 237
Venetian canals, 292–93
Venice, Italy, 293
Verdandi, 103
Veterans, public domain granted to, 250
Vicuña, 145
Vienna, 185
Vireos, 107
Viruses, 17, 238
Visalia, California, 285
Volker, Al, 87–88
Volpe, John, 81, 83, 208

Wallace, Alfred Russel, 15
Walrus, 159
Walt Disney Enterprises, 283–88
Warblers, 107
Washington, D.C., 196, 200
Washington Public Power Supply System, 41
Wasps, 238
Waterfowl, 43
Water hyacinth, 19
Water pollution and water shortage, 49
  from industry, 50
  in Midwest, 59–60
  money for, 60–61
  in New York City, 57–59
  reclaiming water
    desalinization, 63–65
    sewage, 63
  solutions to
    Council on Environmental Quality, 61–62
    Environmental Control Administration, 62
    Environmental Protection Agency, 63
    Environmental Quality Control Council, 61
    Federal Water Pollution Control Administration, 61
  specific waterway pollution, 50–57
Watt, Kenneth, 298–99
Wayburn, Edgar, 261
Weed control, 234
Weeds, 106
Weeks Law (1911), 250
Welland Canal, 19
Wellton-Mohawk Valley, Arizona, 33
West, Irma, 218
WEST (Western Energy Supply & Transmission Associates), 33
West Branch Reservoir, New York, 58
Westchester County, New York, 52, 140
Western Energy Supply & Transmission Associates (WEST), 33
Western plains, 14
Westfall Alternative, 39
West Germany, 60–61
  pesticides and, 217
Westinghouse, 64

West Michigan Environmental Action Council, 216
West Point Military Academy, 53
West Virginia Pulp & Paper Company, 112
Weyerhaeuser (company), 95, 97
Whale oil, 136–37
Whales, 136–37, 296
See also specific whales
White Clouds, Idaho, 265, 268
Whiteface Mountain, 172
Whitefish, 51
White pines, 106–8
White River National Forest, 264
White sturgeon, 41
White-tailed deer, 14
Whitson, Edmund, 150
Whooping crane, 151–52, 281
Whyte, William, *The Last Landscape*, 46, 48, 189
Wild boars, 140
Wild cherry trees, 106
Wilderness, multiple use of, 273–90
Wilderness Preservation Act (1964), 277
Wilderness Society, 38, 42, 83, 118, 278, 281, 293
Wildlife, scarcity of, 135
birds, 140–42
endangered species, 142–63
salmon, 137–40
whales, 136–37
Wildlife Management Institute, 203
Wild orchids, 280
Wild raspberries, 109
Wild turkeys, 78
Williams, Carroll, 236, 238–39
Williams, Jack, 270–71
Willows, 109
Wilson, Alexander, 140
Wind, 105
Wisconsin, 217
Wolverines, 159
Wolves, 21–22, 106, 109, 143, 147, 159, 280
Womble, Martin, 248
Woodchucks, 106
Woodpeckers, 107, 281
Wood storks, 73
Wood warbler, 107
Woodwell, George M., 217, 220–21, 223
Woolly mammoth, 143
World Health Organization, 222–23
Worms, 16
See also specific worms
Wright, Frank Lloyd, 183
Wurster, Charles F., 217, 231
Wyoming, 29

Yannacone, Carol, 294–95
Yannacone, Victor J., Jr., 213, 217, 240, 296
Yaphank-Middle Island Defenders of the Environment, 213
Yarborough, Ralph W., 279
Yellow fever, 232
Yellowstone National Park, 147, 156, 249
Yerba Buena project (San Francisco), 182
Yggdrasil, 103
Yosemite National Park, 256–59
Yucca, 280
Yukon Chinook salmon, 139
Yukon River, 43

Zarchin, Alexander, 64
Zeidler, Othmas, 227
Zentner, Joe, 157
Zeus, 103